Build a Solar Hydrogen Fuel Cell System

by Phillip Hurley

ISBN-13: 978-0-9837847-7-7

Wheelock Mountain Publications
is an imprint of
Good Idea Creative Services
Wheelock VT
USA

First print edition 2013

Notice of Rights

Wheelock Mountain Publications is an imprint of:

Good Idea Creative Services
324 Minister Hill Road
Wheelock VT 05851 USA

ISBN 978-0-9837847-6-0

Library of Congress Control Number: 2013921081

Library of Congress subject headings:

Fuel cells--Design and construction--Handbooks, manuals, etc.
Solar energy--Handbooks, manuals, etc.
Hydrogen as fuel--Handbooks, manuals, etc.
Water--Electrolysis--Handbooks, manuals, etc.

Disclaimer and Warning

The reader of this book assumes complete personal responsibility for the use or misuse of the information contained in this book. The information in this book may not conform to the reader's local safety standards. It is the reader's responsibility to adjust this material to conform to all applicable safety standards after conferring with knowledgeable experts in regard to the application of any of the material given in this book. The publisher and author assume no liability for the use of the material in this book as it is for informational purposes only.

Contents

Electrolyzers

Resources

Templates

Preface

A noteworthy conversation from "Mysterious Island" by Jules Verne, 1874:

It chanced one day that Spilett was led to say, "But now, my dear Cyrus, all this industrial and commercial movement to which you predict a continual advance, does it not run the danger of being sooner or later completely stopped?"

"Stopped! And by what?"

"By the want of coal, which may justly be called the most precious of minerals."

"Yes, the most precious indeed," replied the engineer, "and it would seem that nature wished to prove that it was so by making the diamond, which is simply pure carbon crystallized."

"You don't mean to say, captain," interrupted Pencroft, "that we burn diamonds in our stoves in the shape of coal?"

"No, my friend," replied Harding.

"However," resumed Gideon Spilett, "you do not deny that some day the coal will be entirely consumed?"

"Oh! The veins of coal are still considerable, and the hundred thousand miners who annually extract from them a hundred millions of hundredweights have not nearly exhausted them."

"With the increasing consumption of coal," replied Gideon Spilett, "it can be foreseen that the hundred thousand workmen will soon become two hundred thousand, and that the rate of extraction will be doubled."

"Doubtless; but after the European mines, which will be soon worked more thoroughly with new machines, the American and Australian mines will for a long time yet provide for the consumption in trade."

"For how long a time?" asked the reporter.

"For at least two hundred and fifty or three hundred years."

"That is reassuring for us, but a bad look-out for our great grandchildren!" observed Pencroft.

"They will discover something else," said Herbert.

"It is to be hoped so," answered Spilett, "for without coal there would be no machinery, and without machinery there would be no railways, no steamers, no manufactories, nothing of that which is indispensable to modern civilization!"

"But what will they find?" asked Pencroft. "Can you guess, captain?"

"Nearly, my friend."

"And what will they burn instead of coal?"

"Water," replied Harding.

"Water!" cried Pencroft, "water as fuel for steamers and engines! Water to heat water!"

"Yes, but water decomposed into its primitive elements," replied Cyrus Harding, "and decomposed doubtless, by electricity, which will then have become a powerful and manageable force, for all great discoveries, by some inexplicable laws, appear to agree and become complete at the same time. Yes, my friends, I believe that water will one day be employed as fuel, that hydrogen and oxygen which constitute it, used singly or together, will furnish an inexhaustible source of heat and light, of an intensity of which coal is not capable. Some day the coalrooms of steamers and the tenders of locomotives will instead of coal, be stored with these two condensed gases, which will burn in the furnaces with enormous calorific power. There is, therefore, nothing to fear. As long as the earth is inhabited it will supply the wants of its inhabitants, and there will be no want of either light or heat as long as the productions of the vegetable, mineral or animal kingdoms do not fail us. I believe, then, that when the deposits of coal are exhausted we shall heat and warm ourselves with water. Water will be the coal of the future."

"I should like to see that," observed the sailor.

"You were born too soon, Pencroft," returned Neb, who only took part in the discussion by these words.

Photovoltaic Fuel Cell Systems

The beginning of solar hydrogen technology

The nineteenth century was an exciting time for electrical experimentation and discovery. Shortly after Alessandro Volta demonstrated the voltaic pile to the Royal Society of London in 1800, two experimenters, William Nicholson and Sir Anthony Carlisle, discovered that hydrogen and oxygen could be produced by passing an electric current through water. This was the first demonstration of the principle of electrolysis.

In 1839 in Paris, nineteen year old experimenter Edmund Becquerel discovered the photovoltaic effect when he found that certain materials would produce electricity when exposed to light. In that same year William Grove experimented with reversing the process of electrolysis and invented the first gas battery or fuel cell. In the 21st century, these three discoveries converge in photovoltaic fuel cell system technology.

Primary components

The major components of a photovoltaic fuel cell system are:

1. A photovoltaic power source.

2. An electrolyzer gas source.

3. Fuel cells

The photovoltaic power source can be any commercial BSPM (Battery Specific Photovoltaic Module), or can be an ESPM (Electrolyzer

Specific Photovoltaic Module). The electrolyzer can either be a PEM (Proton Exchange Membrane) or an alkaline type.

Gas processing and storage

Secondary components of the system are a gas processing train and gas storage vessel. The gas processing train can include different components such as scrubbers, catalytic combiners, coalescers, filters and flashback arrestors. Its purpose is to remove particulate contamination, purify the gas stream, and remove excess water or other liquids.

The gas storage system may consist of low pressure storage tanks, bags, tubes, medium pressure vessels and tanks, or high pressure tanks. Other methods of hydrogen storage include liquid phase storage, metal hydride, and nano-tube storage.

Both the gas processing and gas storage system include valves and tube or pipes; plus, depending on the system design, compressors, pressure relief valves, gauges and other equipment specific to the method of storage.

Renewable energy sources and hybrid generating systems

In addition to photovoltaics, other renewable energy sources such as geothermal, tidal, hydro and wind power can be used to produce hydrogen for fuel cells. Any of these sources can also be combined with photovoltaic panels in a hybrid system to power hydrogen production. All of these renewable energy technologies, whether combined or stand alone, have the virtue of no continuing costs of buying and transporting fuel, and thus no dependence on regular supplies of fuels from outside and politically unstable sources.

The choice of a renewable energy source to power a hydrogen fuel cell system will depend on the geography, climate and other resources of the location. I have chosen photovoltaics as the example power source for hydrogen production because it is available to more people than either geothermal, wind, or tidal. For a person without much technical skill, solar power is easier to implement than wind power. It is also less costly up front and not as dependent on location as the other RE power sources noted, so a PV hydrogen system can be portable. In comparison to other renewable energy systems, photovoltaics also has these major advantages: photovoltaic systems are, for the most part, easier to put up and maintain. They have no moving parts to wear out or maintain, so a system is less likely to break down; and they can be gradually enlarged to increase capacity, whereas other systems require greater expense to make upgrades. This does not mean that a photovoltaic system is ultimately less costly, but it does mean that it should be possible to fit a PV system into a more modest budget.

Of course there are exceptions to this. For instance, if you live in Iceland, you would certainly want to explore geothermal energy as your power source. Or, if you live in an area where there is a great deal of cloud cover, wind power or hydropower might be more viable alternatives. Hybrid systems such as a combination of photovoltaics and wind (or any other combination) can be a good a choice for areas where the resources are present but limited. If you are interested in wind power, many wind turbines are suitable for a stand alone power source for hydrogen production, or as a part of a hybrid photovoltaic system.

Solar hydrogen versus solar battery systems

To date, photovoltaics and battery technology have been practically inseparable, combining quite well to convert energy and store it to meet power needs.

However, there are problems with this combination. Spent batteries become toxic waste. At peak power production times the excess energy produced by the PV panels usually gets shunted and lost, because the batteries have a limited charging capacity. For example, where I live in the northern hemisphere there is plenty of sunshine during the summer months to charge my batteries. But, since there is plenty of sunshine, I don't need as much electricity at that time of year. There is less energy demand, but an abundant supply of power. When the batteries are not used as much, it takes less time to charge them, and the photovoltaic panels simply bake in the sun, shunting energy to no purpose.

In the winter the situation is quite different. There is a lack of sunlight, and I need more electricity. The batteries are always hungry for a charge in the winter. This energy inequality can be addressed by adding a wind turbine to a PV system. Fortunately, it is often windy at the times when sunlight is lacking, so these power sources complement each other well. You could also just add more PV panels to your system, which might give you enough power to get through the low sunlight periods, but it would mean that more power is wasted in the times of peak sunlight.

Another possibility is to add a solar hydrogen fuel cell system, which has the potential to operate without interruption because there are no charging limits. During peak power production times, all the available energy that the system can convert can be stored for future use, limited only by the amount of hydrogen storage available.

Time limits for battery storage

Batteries have to be constantly cycled, that is, charged and discharged within a certain amount of time, or they will degrade over time in their energy storage capacity and become less efficient. Hydrogen, on the other hand, can be stored indefinitely for use at a chosen time and season. The carrying capacity of hydrogen as an energy storage medium cannot be matched by present battery technology. This makes a solar hydrogen fuel cell system a good addition to any photovoltaic system.

Hybrid energy storage systems

Some experimenters have combined battery storage and hydrogen storage to see if there are any advantages to such a hybrid storage method. For instance, where BSPMs are set up for normal battery charging duty, a diversion shunt is added so that when the batteries are fully charged, the electricity from the solar panel is diverted to the electrolyzer to produce hydrogen. Simple types of shunt devices that are not particularly energy efficient when solely used for charging in a regular BSPM system are, in this system, quite efficient.

A BSPM system can incorporate hydrogen production in a number of different configurations. A basic hybrid energy storage system requires several 12V panels with an approximate 10 amp output. These panels are connected in parallel with a diversion controller. To operate with the higher voltage and to use it efficiently, the system would include a bank of three electrolyzers connected in series (bipolar connection) that will use four volts each. This will be discussed in more detail at the end of the Electrolyzers chapter.

With a battery and gas storage system, you can have the advantages that each technology provides, and have a base for making the transition to a battery-free hydrogen system in the future.

Solar Panels

Types of solar cells

The basic energy producing unit of a photovoltaic power generating system is the photovoltaic or solar cell. Solar cells can be made from a variety of different materials, however the silicon solar cell is the most common, well developed, and readily available, so it is what we will discuss here.

A silicon solar cell is a solid state semiconductor device that produces DC (direct current) electricity when stimulated by photons. The most readily available types of silicon solar cells are the single crystal cell, the poly crystal cell; and vapor deposition, often called amorphous or thin film cells.

Of these types of silicon cells, the single crystal cell is the most efficient per exposed surface area in producing current. The poly crystalline is the next most efficient, and amorphous is the least efficient per exposed surface area of the group. Single crystal cells are the most expensive to manufacture, polycrystalline comes next as far as manufacturing costs, and the amorphous type is the least expensive to make. This is, of course, reflected in how much the cells cost to the end buyer. In practice poly crystal cells generate slightly less current than single crystal.

There are two types of amorphous (thin film) cells: non-silicon, which can reach efficiencies of 25% using materials such as gallium arsenide, cadmium telluride and copper indium diselenide; and silicon, which at present have efficiencies of 4% to 8% and produce about half the current of a single crystal cell for the same area.

Although silicon amorphous cells are the least costly alternative, they are not a serious consideration for solar hydrogen production as the cost of the added area for panel support structures and extra space needed cancels out some of the savings derived from lower manufacturing and purchasing costs.

Electrical characteristics of solar cells

Each cell, no matter what its size, will produce around .5 volts – some more, some less, depending on the cell. So, if you took a 5" by 5" cell rated at .5 volts and 4 amperes of current, and divided that cell into four separate pieces, each smaller piece would still generate .5 volts. Although the voltage remains the same for each piece, the current output for each would be only about 25% of the original larger section, or about 1 ampere.

This is an important consideration. Using larger area cells in an ESPM allows you to use fewer cells in each string, which saves time and work when constructing panels, as there are fewer tab and bus connects to solder.

Other factors that affect the current output of a cell are the amount of sunlight that stimulates the cell, and the temperature of the cell. The amount of sunlight reaching a solar cell will vary from minute to minute due to particulates in the air, moisture, and cloud cover. Daily variations occur due to the changing position of the sun and the angle of the light striking the cells. These factors and the seasonal angle variations affect the current output of each cell. Heat also affects cell output. When the temperature of the cells rise, the output decreases.

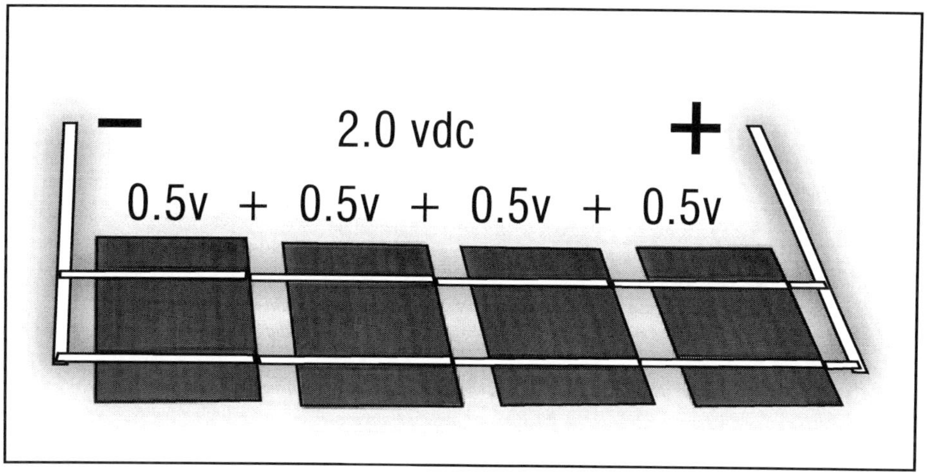

A string of solar cells connected in series. Voltage is added.

Elements of photovoltaic panel construction

A single photovoltaic cell does not produce much voltage and the current output is limited by its size. To augment either the voltage or the current output, solar cells can be connected in either series or parallel. Although it is not always the case, usually the sun-facing surface of the solar cell is negative and the back side is positive.

When cells are connected in series to increase voltage, the negative terminal on the face of one cell is connected with tab ribbon to the positive terminal of the next cell's back. This type of connection adds the voltage of each cell. For instance, for five cells that put out .5 volts apiece and are connected in series, the leads at the top and bottom will give a reading on a voltmeter of 2.5 volts.

For parallel, the faces of the cells are connected to each other and the backs of the cells are connected to each other. With this arrangement, the current output of each cell is added, but the voltage remains the same as the output of one cell. Five cells connected in parallel would give a reading of .5 volts at 20 amperes of current if the cells are 4 amps output.

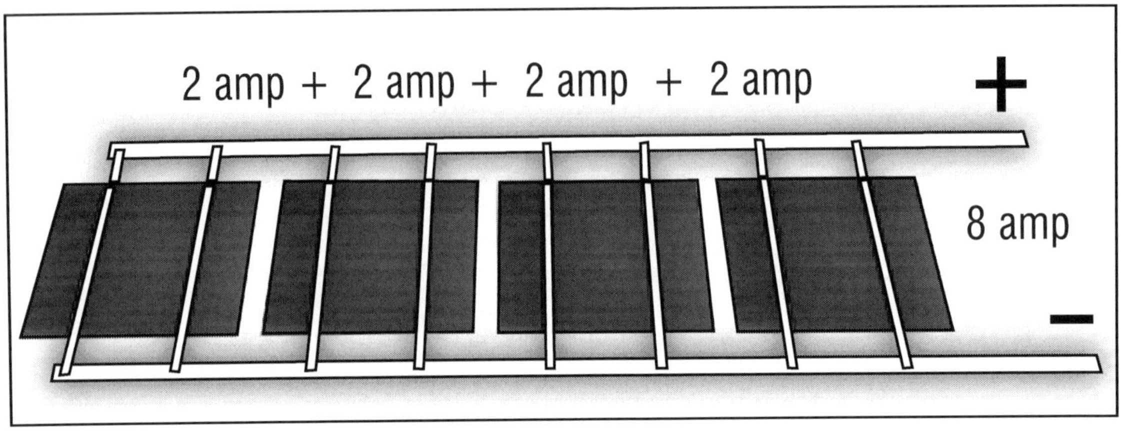

A string of solar cells connected in parallel. Amperage (current) is added.

Connecting two or more cells to each other creates a string. A string is a row of cells connected to each other in either series or parallel. Multiple strings are then connected to each other in series or parallel to form the whole of the solar panel or module.

Two or more solar panels connected together form an array. Arrays can be connected in series, parallel or in series-parallel.

Task specific photovoltaic modules

All photovoltaic modules are classed as some form of TSPM (Task Specific Photovoltaic Module). This means that every solar panel is designed to perform a particular task.

Conventional solar panels are called BSPMs (Battery Specific Photovoltaic Modules). Most panels sold commercially are of this type and are designed solely to charge battery systems. These panels come in 12 volt, 24 volt and 48 volt configurations. Most have short circuit current ratings of from 2 to 10 amperes.

ESPMs (Electrolyzer Specific Photovoltaic Modules) are uniquely designed to match the power requirements of electrolyzer systems.

This makes them more energy efficient and economical for their intended use. Whereas BSPMs output higher voltages and lower currents, ESPMs output higher current and lower voltages.

Although not always possible, the best engineering practice is to design any power supply to match the specific appliance that will use that power, so that as little energy as possible is wasted. Still, BSPMs can be used efficiently for hydrogen production if they are used to power a bank of electrolyzers rather than just one.

For instance, a two panel BSPM array connected in parallel will sufficiently power three electrolyzers connected in series. For more current you would add more panels in parallel to the array.

Another option is to insert a voltage divider, or DC to DC converter between the BSPM and electrolyzer, which will give the correct current and voltage for the electrolyzer. However, with this configuration there will be some power loss. If that is not a concern for you, then it is a viable option.

Basic components to build a solar panel

- Solar cells
- Bus ribbon wire
- Tab ribbon wire
- Sheet metal backing (or other rigid material)
- Support frame
- Cover (Plexiglas®)
- Electrically insulating underlay (if you use a metal backing)
- Screws, bolts
- Power distribution box
- Wire and terminal blocks
- Schottky power diodes with heat sinks (optional)

Designing the ESPM

Since solar cells come in a variety of shapes and sizes, it is not possible to give exact dimensions for the particular panel you will build. Your panel dimensions will depend on what size cells you are going to use, how much power it will generate, and where it is going to be situated. These factors dictate the final dimensions and the type of materials used for your particular panel.

Front and back views of the finished PV panels

I will show you how a panel is designed and constructed, so that you can make your own calculations based on the materials you have and your power demands.

The first design consideration is to determine what the power needs of the electrolyzer will be.

I decided on a photo-voltaic power supply that would provide somewhere between 10 to 20 amps, at 4 to 6 volts. This would be sufficient to power my electrolyzer.

Choosing PV cells

The next choice is the type of solar cells to use. Cells come in a wide variety of shapes and sizes. The most frequently encountered shapes are square, pseudo square (square with angled corners) and

Various solar cell shapes. Top left to right, rectangle, pseudo-square; three round cells on the right; and bottom left, two square cells.

round. I chose a pseudo square shaped single crystal cell that was rated at 4 amps at .55 volt per cell. These cells were purchased as cosmetically blemished and off spec (sub standard) thus, they were available at a reduced price. Each cell is 5" in length and width.

When shopping for cells, you can purchase new cells with no flaws or new cells with flaws, either cosmetic or those that are off specification. Off spec cells are cells that did not have the expected output when tested at the factory. They are still good cells, but may not provide the current and voltage that would make them suitable for a commercial panel. When considering off spec cells, keep in mind that the lowest output cell on a panel is what the panel's final output will be. So, be sure that the lowest output cell is within your acceptable limits.

Cosmetic flaws can be anything from chips off the sides or corners, discoloration, or lack of reflective coating. Cells can be just cosmetically flawed and putting out full output; or they can be cosmetically flawed and off spec. For instance, a lack of reflective coating on parts of a cell (cosmetic blemish) can reduce their output as they will reflect more and not absorb as much light.

Higher current cells are larger and cost more per cell. However, larger cells mean fewer tab and bus connections to make, and they reduce the number of cells needed to produce a given amount of current.

I have listed solar cell suppliers in the resources section, but you can also call major solar cell manufacturers. Request their prices for cosmetically flawed and off spec cells, and find out what their minimum order is. Popular internet auction sites are also sources for cells. Take your time, shop around, compare prices, and consider customer service.

Testing solar cells

Each cell should be tested for voltage and current output before you solder the tab ribbon to the cells. Even if you are using up-to-spec new cells, test them before you solder. They might still be flawed or have been damaged in transit and handling.

Solar cell manufacturers use an array of lights that closely mimic solar output to test cells. This gives excellent results, but is expensive and not really necessary.

The most low cost test option is simply to take the cells outdoors on a very sunny day with a multimeter. The good news is that the sunlight testing station is free. The bad news is that if it is not a perfectly clear day, barely perceptible changes in light intensity occur within seconds due to a variety of atmospheric changes. This can easily throw your readings off a bit. However, this is not a really major concern – ball parking it should be sufficient. A fairly decent sunny day works for this technique.

With the multimeter, measure the open circuit voltage and short circuit current for each cell. Write down the reading for each one. The cells do not have to exactly match each other in voltage and current output. The point is to match the cells so that all of the cells put out voltage and current at or above what your target output is. The lowest single cell output will limit all the others to its output level, so try to match them as closely as possible.

When I was testing the cells for this project, I waited for the first sunny day to get out and test the cells. I tested and selected my cells, and connected them together in a string, The next day, when I performed the second test, I was surprised to find that my current readings were much higher than I had expected. What I thought was a very sunny day for the first test was not as sunny as the day of the second test. I realized that there must have been more particulates and or moisture in the air on the first test day than on the second test day.

The time of the day, season of the year, and atmospheric particulates and moisture will cause changes in the readings. Full output can be tested best in the summer at high noon on a clear day. This doesn't mean that cells cannot be tested this way at other seasons of the year or times of the day. Just take into account that output variations can occur because of these factors. You will, of course, get lower output readings in the winter than in the summer as the light has more atmosphere to traverse due to the tilt of the earth. Other times than high noon or when the sun is at zenith in your location will give you a lower reading. Despite all these variances, I find this to be an excellent method of testing cells.

If you test cells this way, make some sort of holder that can be tilted at an angle, and that will hold the cell so that you can easily take readings. You should be able to adjust the direction and angle of the device so that it faces the sun as directly as possible. Be sure that when testing that you do not shade part of the cell – it will affect the reading.

Laying a cell on a piece of copper clad circuit board is helpful for taking readings. If you do use a copper clad circuit board, make sure the surface is clean and that the back of the cell is touching the copper clad well. Lack of good contact will give you a false or weak reading.

When you take your cell reading for either open circuit voltage or short circuit current, touch one probe to the solder finger on the top side, and the other probe to the copper clad board that is in contact with the solder finger on the other side. To measure voltage with a multimeter you simply dial in "voltage," place the probes on either end, and note the voltage. To measure current move the dial to current measurement and note the reading. For this project you will need a multimeter that can measure at least 20 amps. You could also use a separate voltage meter and ammeter.

Another way to test cells is to take one cell and expose it to sunlight on a very clear day, then test this same cell under an artificial light such as an ELH projector lamp bulb which has similar characteristics to daylight; and/or a daylight photoflood such as a BCA-B1 which has a daylight color temperature of about 4800° Kelvin. Natural daylight has a Kelvin color temperature that varies between 5000° to 6000° Kelvin.

Or, take a reading for one cell (this will be your control) under the artificial light at a specific distance. Then, take the other solar cells and see if they are close in output to the control cell. This will give a relative comparison to a cell which you know has the output that you need.

You can also purchase a calibrated solar cell. My book *Build Your Own Solar Panel* discusses in detail cell testing with a calibrated cell.

Solar cell manufacturers test their cells using a xenon light source with filters, under "AM1 conditions." An AM1 condition is when the sun is directly overhead on a very clear day. AM stands for air mass. The number indicates the amount of air that sunlight has to travel through, and the resulting spectrum and intensity change due to the variations in air mass. The change in spectrum is visible when you see more red during sunrise and sunset than at midday. The sunlight has more air mass to go through at sunrise and sunset, so the Kelvin color temperature will be different at those times than when the sun is directly overhead. In AM1 conditions, the irradience affecting a surface is considered to be about 1000W/m2 and is called one sun or full sun.

This irradience or insolation (a term that is a shortened version of "incoming solar radiation") figure is used to size photovoltaic systems. Of course, the amount of incoming solar radiation depends on a number of factors such as cloud cover, moisture, and particulate concentration in the atmosphere. Each geographic region has its particular climate characteristics to be considered when calculating the number of panels required for the photovoltaic fuel cell system.

Tab and bus ribbon

When you have decided what type of cell to use for your project, you can move on to purchasing the tab and bus ribbon that will connect the cells and strings of cells. Tab and bus ribbon are made from soft copper that is pressed into a flat wire. This ribbon is tin coated to make it easier to solder. Most tab and bus wire used for BSPMs is .003", .004", and .005" thick.

For ESPMs, a .005 thick tab ribbon is sufficient to connect the individual cells to the bus ribbon, but the bus ribbon needs to be wider to accommodate the greater current carried. For the panels described here, a 20 amp capacity conductor is needed for the bus wire. Most suppliers of cells do not have bus ribbon in the thicknesses required. Enquire and see what is available.

Tab and bus ribbon can be purchased in the thicknesses required and cut to any width from manufacturers, such as E. Jordan Brookes Co. They cut the ribbon from larger rolls and can provide any width and thickness of tinned wire desirable. Most tab ribbon widths supplied by secondary cell dealers usually run about 1/16" to 1/8", and bus ribbon is usually cut at 3/16" widths. Measure the width of the

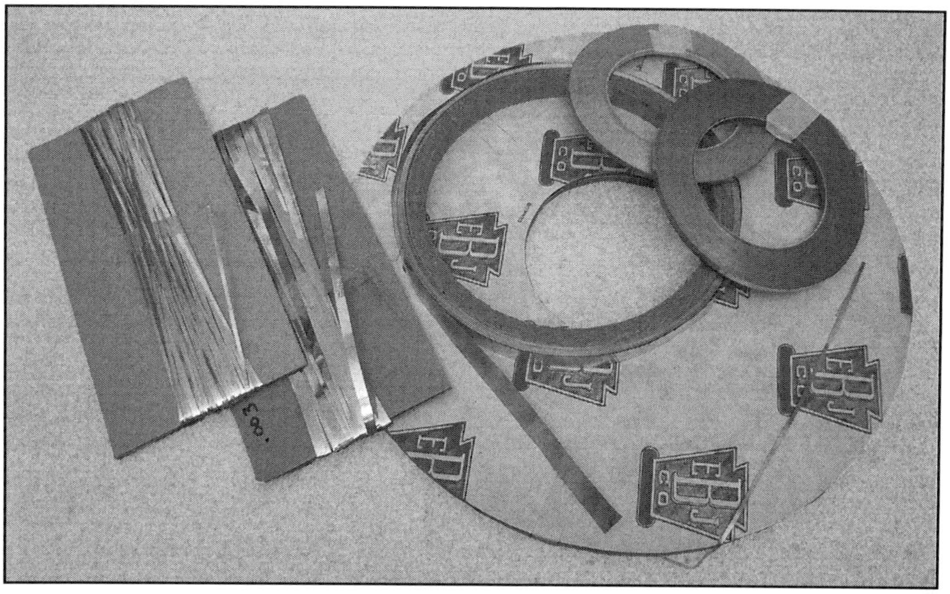

Tab and bus ribbon

solder fingers on your cells to see if you can use a larger width cell connect tab wire. The more surface you cover, the better. These sizes are generally adequate for most cells and solar panels. The $^{1}/_{16}$" width is usually fine for most cells. Ask your dealer what thickness and width the tab ribbon is.

For an ESPM as discussed in this book, use at least a .005 thick ribbon for the cell connects. The bus ribbon needs to be able to carry the current of this ESPM, for instance .010" thick, at $^{3}/_{8}$" wide; or a .020" thick at $^{7}/_{32}$" wide.

You can make your own tab and bus wire. Copper foil is available in thicknesses from .002 to .021 from McMaster-Carr or other suppliers. It can be cut into thin strips to your particular specifications. You can also use flat grounding braid for bus ribbon, as it is made for heavier current carrying capacity. Grounding braid comes in a variety of thicknesses and widths and is usually tinned. Make sure to use Tinnit® to tin the tab and bus if you make your own – it also helps to minimize oxidation. You can use round wire, either solid or stranded, for bus connections as well as cell connects, if you desire, but the flat ribbon is probably easier to work with. See my *Build Your Own Solar Panel* for more details.

Square mils

It is easy to find the current capacity of round wire in a variety of published tables for the photovoltaic enthusiast, but it is not so easy to find the current capacity of ribbon wire. Not very many people engineer their own PV panels.

Although the current carrying capacity of a wire is based on factors other than wire diameter (type of insulation, stranded or solid), wire gauge is a general indicator, within limits, of the actual current allowable for a certain area of a conductor.

If you are working with a conductor that is flat and of a certain thickness and width, it is useful to be able to compare its current carrying capacity to round wire conductors.

Wire gauge is usually indicated in what is called AWG or B&S. AWG stands for American Wire Gauge, and B&S stands for Brown and Sharpe gauge. They are one and the same. In the table at right, wire gauges are listed with their rated amp carrying capacity and their area in circular mils and square mils.

Since you will be building a panel that delivers 20 amperes of short circuit current, use the table (next page) to see what size wire carries that current adequately. Note that #14 gauge will carry 20 amperes, but you now need to know what size of flat ribbon wire will carry that same amount of current.

Circular to square wire conversion table

Size AWG	Current carrying capacity/amps	Resistance/ ohms per ft.	Circular mils	Square mil equivalent
14	20	0.0026	4,107	3,226
12	30	0.0016	6,530	5,128
10	35	0.0010	10,380	8,152
8	50	0.0006	16,510	12,967
6	70	0.0004	26,250	20,617
4	90	0.0003	41,740	32,783
2	125	0.0002	66,370	52,127
1	150	0.0001	83,690	65,730
0	200	0.0001	105,500	82,860

Values are for copper conductors only.

Square mils/current capacity

One square mil equals .001". To find the current carrying capacity of square or rectangular shaped conductors such as bus ribbon, multiply the thickness in inches times the width in inches. The result will be the square mils of the conductor. The table on the previous page gives the carrying capacity based on square mils.

Voltage drop

To find the voltage drop for any length of wire run, multiply the resistance per foot of the particular size conductor, times the number of feet in the run, then multiply this times the current (amps) you will be running through the conductor. The result is the voltage drop. For example, an ESPM with an output of 16 amps at 4 volts and a wire run of 10' to the electrolyzer through a #10 conductor, will give a voltage drop of 0.16v. This means that at the terminals of the electrolyzer the system will be able to deliver 3.8v.

R x Ft. x Amps (resistance x length of run in feet x amps) = voltage drop.

To get the current carrying capacity of the ribbon wire, simply multiply the thickness in inches by the width in inches. This will give the area in square mils. Take this square mil figure and match it up with one of the wire gauges listed in the table on the preceding page. Then, multiply your square mil figure by .7854 to give the area in circular mils. Whatever gauge the ribbon is close to in circular mils will tell you generally what the current carrying capacity of the bus ribbon is. Note that this table is only correct for copper conductors.

As an example, consider a bus ribbon .020" thick and 7/32" wide. Convert 7/32" into its decimal equivalent (.21875) and move the decimal place to 218.75 because 1 mil = .001". Then multiply 218.75 times 20, which gives you 4375 square mils. Either check the table or convert these square mils to circular mils by dividing 4375 by .7854, which is 5570 circular mils. As you can see from looking at the conversion table, this figure falls between a 14 and 12 gauge, and any wire size within this range would be sufficient to carry 20 amps.

Building the ESPM

Tool list

All available at local hardware or electronics stores.

Screwdriver

Hacksaw, and other types of saws to cut metal and other items

Ruler, T square, other measuring devices as needed

Exacto knife, razor knife, other cutting instruments

Caulking gun

Soldering iron, 60 watt, with screwdriver-type tip

Paint brush to spread caulk

Drill, either hand or drill press

Multimeter that can measure current

Materials list for two panels

All should be available at local hardware or electronics stores, unless otherwise noted. See also Suppliers List, page 201

40 PV cells, Plastecs or other

Tab and bus ribbon, E. Jordan Brooks or other

Fiberboard 28"x 30"

Clear silicone rubber caulking

20 Stand-offs to hold Plexiglas® in place, 1/4" wide, rubber or plastic

Silicone solid rubber, 1/32" thick, 12"x 36", 30A durometer, McMaster-Carr, part #8622K31

Tinnit® electroless bright tin plate, All Electronics, part #ER-18

16 Screws, 10-24 1 1/2 " long, and nuts to fit (both stainless steel)

2 Aluminum sheets about 1/16" thick: 28"x 30" Local hardware store, local metal supplier, sheet metal shop, or McMaster-Carr

0Materials list for two panels (continued)

Wire – red and black zip cord. All Electronics, automotive store. All Electronics part #WRB-10 (10ga.) or WRB-12 (12ga.)

2 Wire connectors, crimp on type

2 Junction boxes, plastic case, 4.7"x 2.6"x 1.55" All Electronics, part #1591-CSBK

2 Terminal strips, 2 position, dual row, All Electronics, part #TS-250 or equivalent

1 Terminal strip, 4 position, dual row, All Electronics part #TS 6034 or equivalent

Electrical tape, liquid electrical tape, liquid rubber or shrink tube

2 Sheets nylon screen 28"x 30"

2 Sheets Plexiglas® about 3/32" thick, 28"x 30", McMaster-Carr or hardware store

Aluminum bar stock .25" thick, by 1" wide, hardware store, local metal supplier, sheet metal shop, McMaster-Carr

Solder – can be 2-4% silver solder or regular 60-40 depending on what works for your particular PV cells. 2-4% silver is recommended although it may not be necessary. Do not use acid flux!! Use rosin.

Flux pen, 2 pens, or as needed. Do not use acid flux. HMC Electronics, #186FP Mildly activated rosin, type RMA

Crazy glue or epoxy

2Schottky diodes, 45prv,75 amp, Surplus Sales of Nebraska, part #SDI-USD5096F (optional). 2 Heat sinks for diodes, Surplus Sales of Nebraska, #HSK-HEATCNK90 (optional).Distribution box to house diodes and heat sink (optional)

Grommets for junction box and distribution box (optional)

Building the ESPMs

Two ESPMs were constructed for this project. The design was based on materials that were on hand, so the finished products are not optimal, but were relatively inexpensive. For each panel back, I used a 28" x 30" sheet of rigid aluminum metal that was a little less than 1/16" thick. I purchased a piece of fiberboard and cut it to 28" x 30" (the same dimensions as planned for the finished panels) so that I could use it as a "peel," a soldering platform, and layout grid

The overall layout for the solar panels, showing placement of rubber spacers

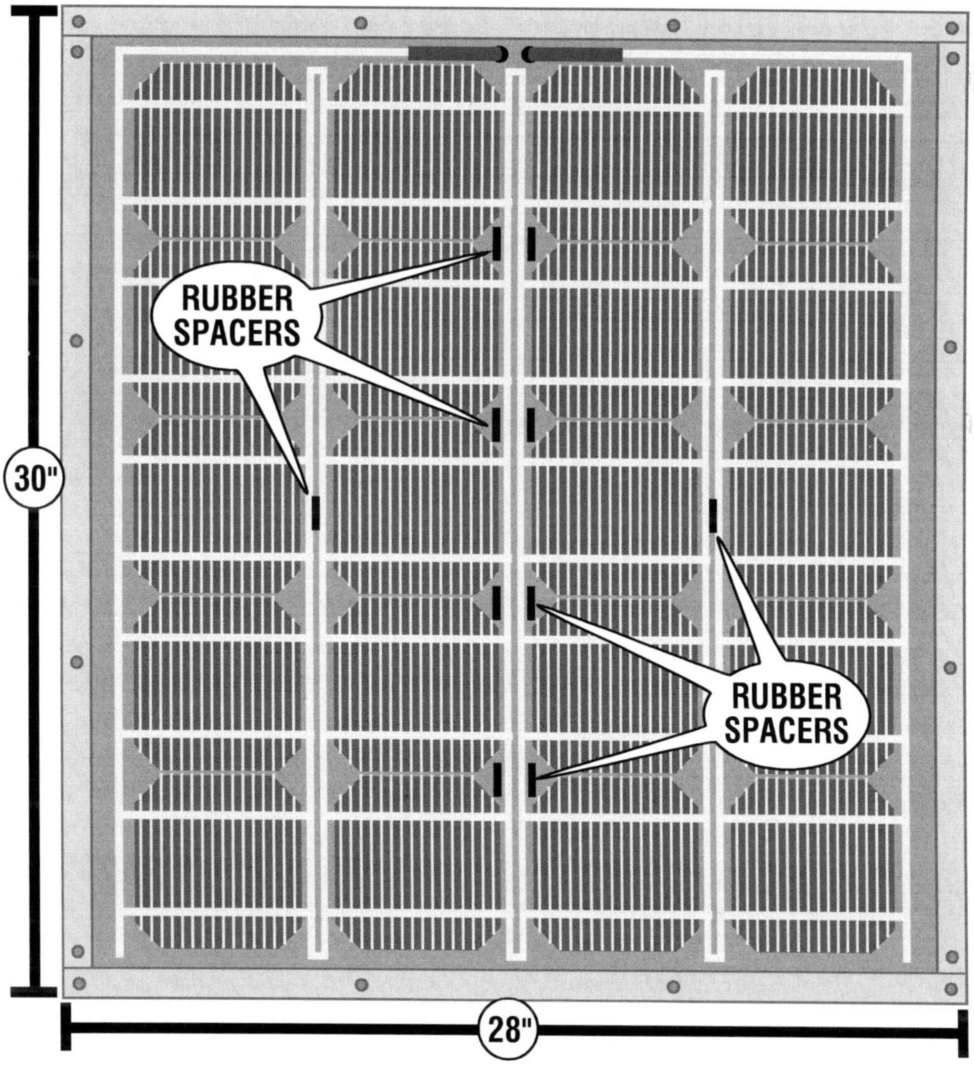

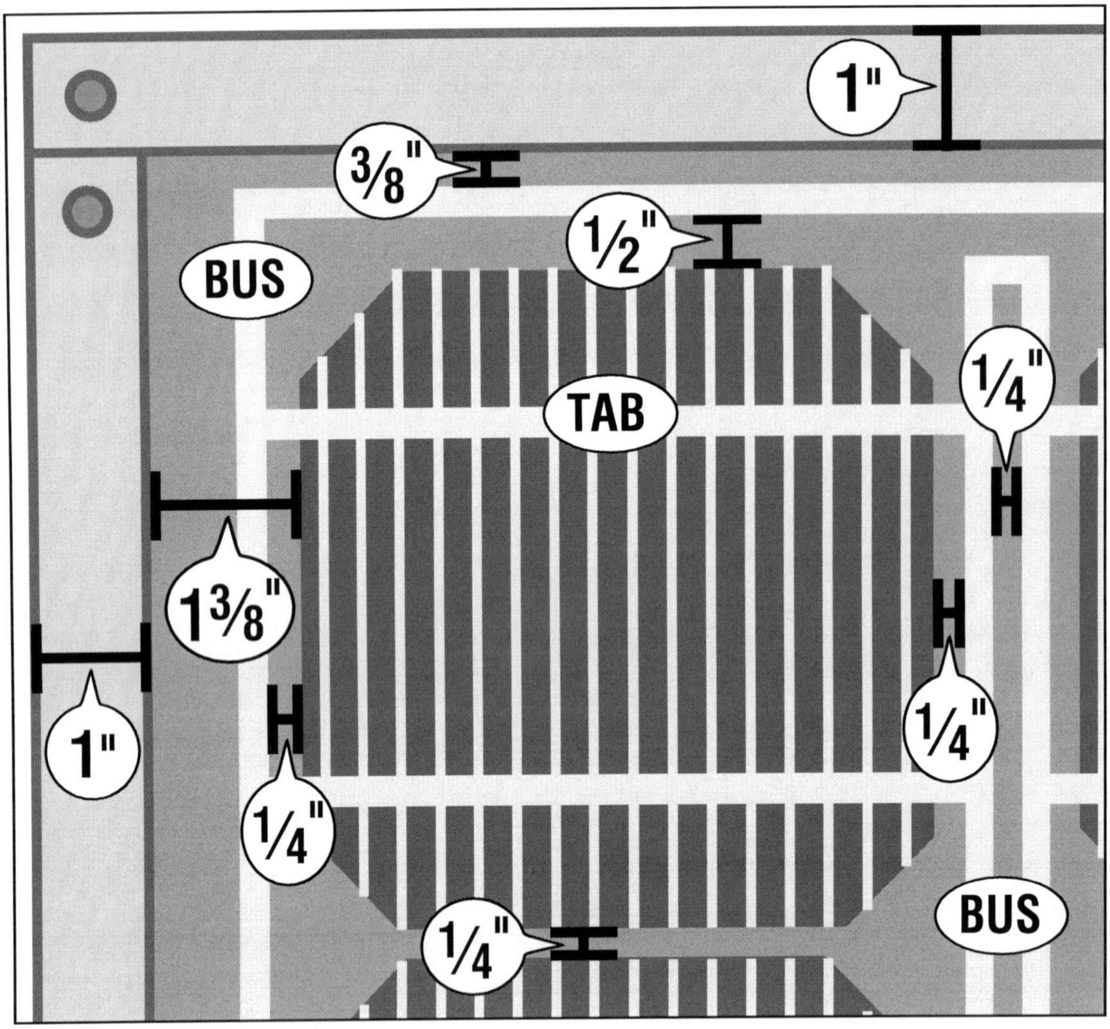

Detail of spacing for cells, bus and tab wire

for the cells, tab and bus ribbon positions for the panels. I used 4 pieces of 1/4" x 1" aluminum bar stock for a frame/support and as a spacer between the back of the panel and the cover. For the covers I used 28" x 30" x 3/32" Plexiglas® sheets.

With a pencil I outlined an inch border on all four sides of the fiberboard. This one inch border marks the space taken up by the aluminum side bars which form the frame of the panel.

The particular cells that I had were 5" in width and height. I laid out the cells on the fiberboard to see how many I could fit in the 26" x 28" space with enough space left for tab and bus ribbon connects as well as enough space to stay comfortably away from the metal edges. With these cells I could just fit twenty cells (4 strings of 5 cells each) with sufficient space for tab and bus ribbon connects, and power take-off leads.

After calculating what looked like a pleasing and practical configuration, I outlined every part that would be laid on the board. This outline was my guide for soldering and connecting the cells and strings of cells.

Cutting the tab ribbon

The individual cells have two tab finger lines on each side of the cell, so four tab ribbons had to be cut for each cell. The tab ribbon runs the total length of the cell, so the length for each tab would have to be 5", plus an extra 1/4" for space between the cell and the bus ribbon connect, plus the width of the bus ribbon that the tab is connecting on to. So, I needed to cut 160 pieces of 5 7/16" long tab ribbon for the 40 cells that would be in the two panels.

Crimping the tabs

For these panels, I did not crimp the tabs, however, you should crimp the tabs so that your connections can expand and contract. Panels are usually exposed to environments with wide temperature fluctuations, from very hot to very cold. Crimping permits slight movement and gives flexibility that helps to ensure that the tabs stay connected to the cells and bus ribbon. Temperature extremes cause expansion and contraction that can break the connections over a period of time.

If you add a crimp to the cell tabs, add another 1/4" to the 5 7/16" length mentioned above. The 1/4" will allow a 1/8" high crimp. It can be a little less if you wish, but the crest or raised part of the crimp should be centered so that it falls mid point between the cell and the bus ribbon. There is more detail about crimping tabs in my book *Build Your Own Solar Panel.*

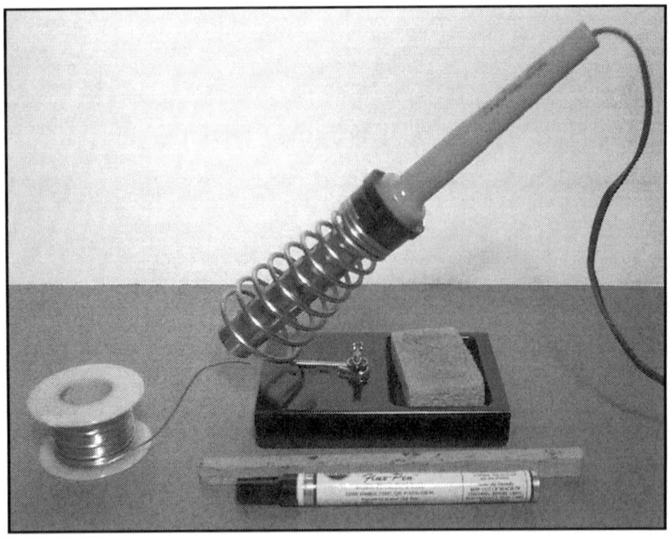

Soldering tools

Tinning the tabs

Tinning is a simple process in which you touch the tip of the soldering iron with the solder wire, and cover it very liberally with solder. The solder melts on the tip and forms a glob which can then be deposited on the surface of the tab ribbon as you run the iron up and down the length of it. Coat the ribbon wire completely and generously for the length that is needed.

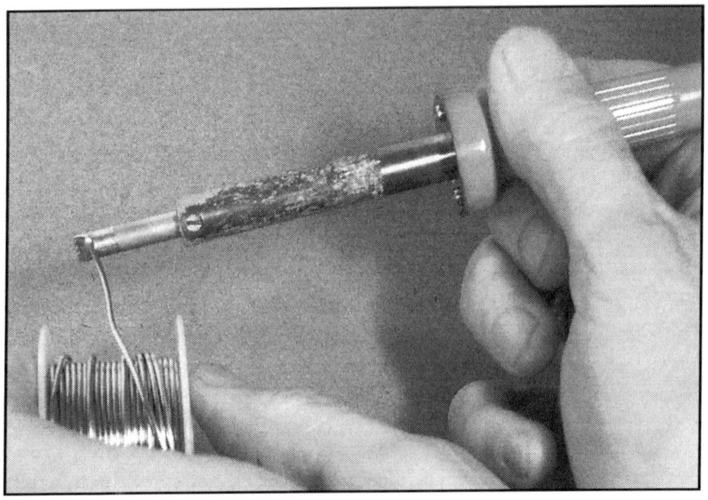

Melt some solder on the tip of the hot soldering iron

The purpose of tinning is to make a solder base from which to form a joint when the ribbon is attached to the cell fingers

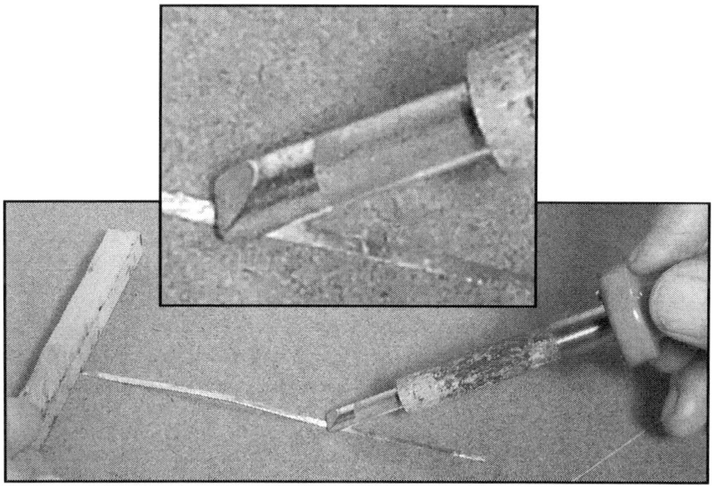

Apply the tinning to the tab ribbon

or to tab or bus ribbon. Although the tab and bus wire comes already coated with a thin layer of tin, it is still necessary to apply more tinning on the areas to be joined.

Practice tinning a few pieces of tab or bus ribbon to get the knack of it. If you miss areas that should be tinned, the joints might not be as solid and strong as you will want them to be. Try to coat as evenly as possible by running the iron down the length of the ribbon. If you have not finished the entire length and the solder is thinning out too much, put more solder on the tip and continue until the surface is fully and evenly coated.

Each tab was tinned from one end for a five inch length on one side (to be connected to the cell); then, on the same side, but on the opposite end, a 3/16" length was tinned (to be connected to the bus ribbon).

This can be done according to how you want the finished product to look. If you want the bus ribbon to cover all the tab ends when you look at them from the front of the cells, then tin the 3/16" on the same side for half of them (for one panel that would be forty) and 3/16" on the opposite side for the next forty (see illustration next page). If you do this, when the bus ribbon is laid down to be soldered to the tabs, all the tab ends will be under the bus ribbon. It gives a neater finished look to the panel, but is not necessary and will not affect performance.

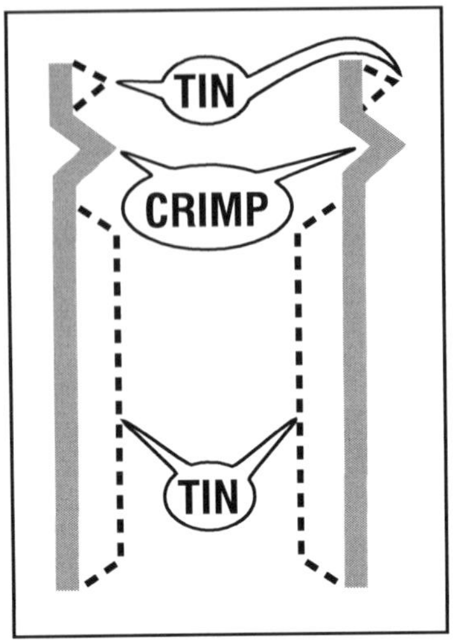

Tinning crimped tabs

If you have crimped the tab ribbon, be sure not to get solder on or in the crimp as this will make it rigid and inflexible. This would negate the purpose of the crimp, which is to add flexibility to expand and contract with temperature changes.

Cells from different manufacturers have different kinds of solder fingers on the faces and backs of the cell. Some of the older types of cells that you may run across have the back of the cell completely coated with solder, and the fronts have thick solder fingers. Newer types of cells usually have the metal fingers put on by a vapor deposition technique, or they are silk screened on. The conductive metal in these deposited or screened fingers can be nickel, silver or other conductive alloy. Because of the various methods of deposition and the nature of the conductive metal used, be sure the soldering materials you use is compatible with the solder fingers on your PV cells.

Generally, use a solder with a 2% to 4% silver content. A 96/4 (96% tin, 4% silver) silver solder is generally available at most electronic stores. The melting point of 96/4 is about 460°F. However, I found that I could use regular 60/40 (60% tin, 40% lead) on the cells I had. I applied solder to the tabs liberally during tinning, and it soldered quite well to the cell fingers. 60/40 is less expensive and if it works well on your cells, use it. The melting point of 60/40 is about 430°F, which is a little lower than the tin/silver alloy.

When you purchase cells, ask which type of solder will work the best. If the recommended solder is a silver type solder, try it and see how well it works, and then try some regular 60/40 and compare.

Tabs will tend to pull off easily after soldering if you do not have enough solder tinned to the surface of the ribbon to make the joint.

Fluxing the cell fingers

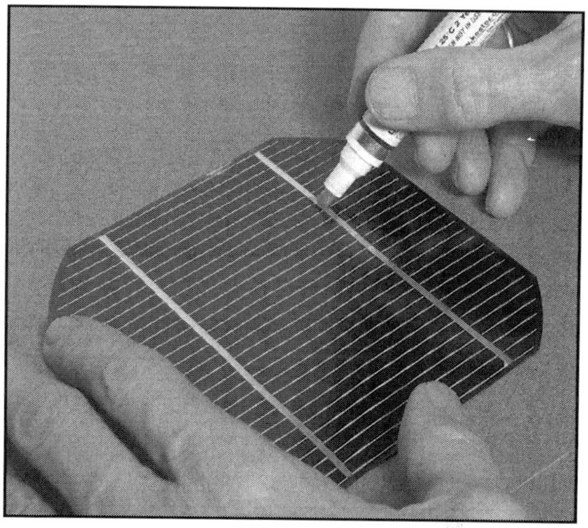

Applying flux to the cell fingers

The secret to a good bond, especially when working with solar cells, is using the right flux and using it correctly. Flux cleans the metal surface and reduces surface tension between the solder and the metal it will adhere to. This is important to make a good electrical contact. Do not bother with paste or liquid flux. Use a flux pen as indicated in the materials list. The pen will put the right amount of flux down and do it very quickly and easily when working with cell fingers. It does not waste flux, and one pen was enough for this two panel project. The tip of the pen is perfect for cell tabbing. Flux pens have a felt tip which is spring loaded with flux when you push the tip against a surface. When working with solar cells, do not load the felt tip while the tip is on the cell. Push the tip against a surface other than the cell. This will load the tip with liquid flux. Then, take the tip and run it down the cell finger. It will cover the entire finger with flux. Do not press anything against the cells – they are fragile and liable to crack. Using a flux pen rather than other types of applicators is very fast, easy and efficient.

Apply the flux to one tab finger at a time; that is, flux a finger and then immediately solder a tab to that finger, Then, move on to the next finger, flux, and tab, and so on.

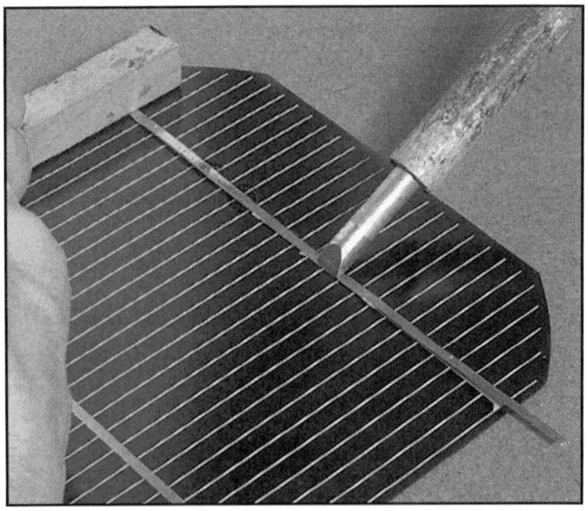

Soldering tab ribbon to the cell front (above) and the cell back (below)

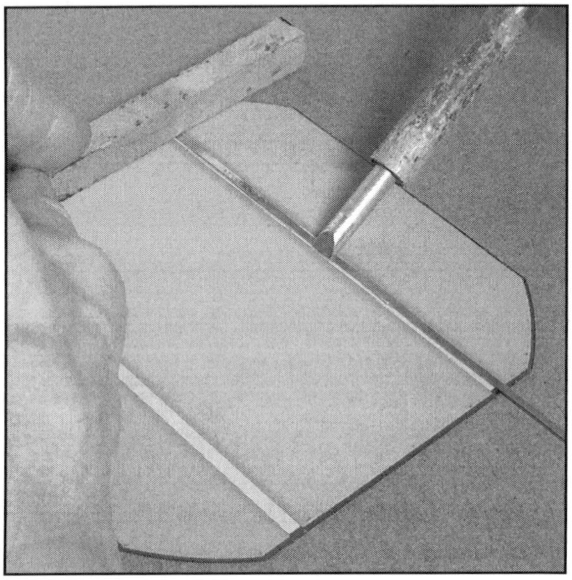

Soldering the tab ribbon to the fingers is easy. Position the tab on top of the finger and hold one end with a stick or other object to keep it steady and aligned while you solder. Then move the iron down the length of the tab. You will see the solder melting as you move. Do not leave the iron stationary in one place too long, as the heat from the iron can damage the cell. When moving the iron along the tab ribbon, don't move too slowly or too fast, but just the right speed to melt the solder along the length as you proceed. Practice on some scrap cells to get a feel for the timing.

Making a parallel connected string

After all of the cells needed are tabbed, connect them into strings, according to your design. My panels called for four strings of five cells each. Position the tabbed cells within the cell outlines drawn on the fiberboard, and tape each cell by the tab ribbon to the fiberboard so that the cells will not move. Do not put tape over the surface of the cells as this can leave a gummy residue, and when you take the tape off, it can crack the cell.

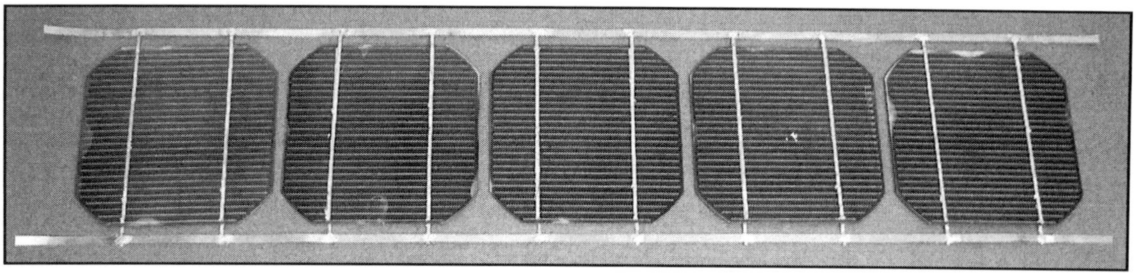

String of PV cells connected in parallel

Next, cut eight pieces of 26" long bus ribbon and lay them out in the bus ribbon position drawn on the fiberboard, over the tabs which extend from the cells. Tin two pieces of bus ribbon at the points where the tab ribbons will connect with them, and then tape the bus ribbons in their positions on the board on either side of the cells. Although the tape is not necessary, it's helpful to keep everything steady while you are soldering. Tin the iron lightly, and apply the iron to each tab solder point – eight soldering points on each side of each bus. This makes a 0.5 volt 20 amp string with a negative bus on one side and a positive bus on the other side.

Solder the tabs to the bus wire to connect the cells in a string. Note that this is a positive bus wire, connecting to the PV cell backs.

Trim the tab ends that extend beyond the bus wire. Note that this is a negative bus wire, connected to the PV cell faces.

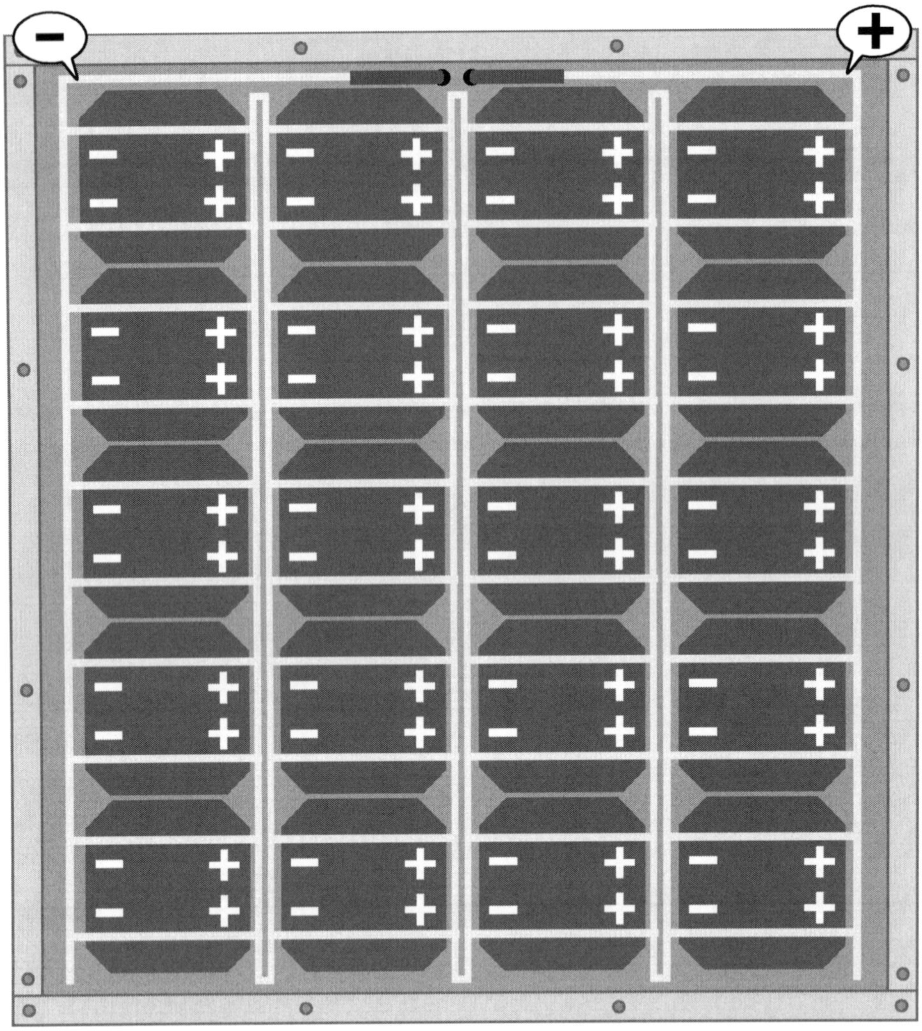

Panel layout showing electrical connections

Test the short circuit current under sun light by putting the multimeter probes on the ends of the negative and positive bus leads, then test the voltage. If you are not getting the reading you should be getting, check the solder joints and resolder if necessary. Solder the other strings in the same manner on the guidelines of the fiberboard and you will then be ready to connect each of these tested parallel connected strings of cells in series to each other.

Soldering the strings together in series.

Connecting the strings in series

With the strings laid out in the final positions they will occupy in the panel, connect the strings with bus ribbon. For my panel, I cut 18 short pieces of bus ribbon the length needed to connect the string buses that lie next to each other (see illustration above and on next page). Tin $^{3}/_{16}$" on each end of the short pieces of bus ribbon.

These strings will be connected to each other in series, which means that the positive bottom side of each string is connected to the negative top side of the next string, and so on. The bus bar on the left side of the first string on the far left will be negative. The bus bar on the right side of this first string will be positive. This positive bus is connected to the negative bus of the string next to it and so on. Make sure that when the strings are laid out on the board they are all aligned the same way, for instance, the left bus of each string is negative, and the right bus of each string is positive. With this particular arrangement, the negative bus ribbon that will connect to the take-off will be on the left edge of the panel, and the positive bus ribbon that will connect to the take-off will be on the right edge of the panel.

Be sure that the layout is correct and that everything is taped down.

Soldering the connections

Solder the buses together with the small pieces of bus ribbon interconnects. Face the tinned side down on the bus ribbon and apply the iron on each end while securing the other end with a stick or other tool so that it doesn't move when you solder (see illustration below). When the string buses are all connected, test the voltage and current again.

Connect the strings with small pieces of bus wire.

Test outdoors

Trim any excess bus wire when the strings are connected

With everything still taped to the fiberboard, take the layout outdoors and tilt the board toward the sun at an angle to give it maximum sun. Take the open circuit voltage reading by placing the one multimeter probe on the far left bus ribbon, and the other probe on the far right bus ribbon at the top of the strings. This is where the power take-off bus will connect with the strings. This reading indicates the total voltage, which should be the sum of .5 volts per string. In the case of the panels that I built, this was around 2.0 volts.

Next, measure the current output by changing the setting on the multimeter to test current. Again place one probe on the far left bus ribbon and the other probe on the far right bus ribbon. This gives the short circuit current reading. The panel I built gave a reading of around 20 amps. You may get a lower reading depending on the time of day and season, etc.; but the reading should be somewhere from 16 to 20 amps. As mentioned before, it is best to wait for a clear and bright sunny day and go out about noon when the sun is highest in the sky. Make sure the fiberboard is tilted to take full advantage of the sunlight. This will give your maximum reading.

If you are not getting the voltage or current readings that you should, visually inspect the cells and the connections. If you cannot see a flaw, try probing with the multimeter and take readings on the strings and cells at different locations to try to find the problem. If one of the cells is cracked, replace it by applying the soldering iron to the joints to desolder and remove it from the string, then replace it with another cell.

Usually you will not encounter any problems in the final test, if you test as you proceed with assembly. The first test should be on each cell individually to check general output. The second test occurs when you finish each string of cells to make sure that the string's output is what it should be. The third test is to take readings when all the strings are connected to each other. This tells you the final output value of the panel.

Handle cells and cell assemblies gently

During the construction process be sure to handle and move the cells carefully. I usually have several pieces of fiberboard on hand so that I can peel/slide the cells from one board to the other very carefully and quickly when I need to make space while I am soldering. If you pull or stress the ends of the cell tab connects, you can inadvertently pull the tabs off the cell. The keyword here is gentle! If one of the soldering joints does come loose, resolder as needed.

Building the panel structure

When the strings of cells have all been connected, you are ready to assemble the rest of the panel.

For the ESPMs I made, I used precut 28"x 30" sheet aluminum that was about 1/16" of an thick. For the side bar/framing material I used 1" by 1/4" aluminum bar stock cut into four 28" pieces for each panel. For the covers I used 3/32" thick Plexiglas® that was cut at the hardware store to 28"x 30". When you purchase a cut Plexiglas® sheet, it usually has a thin film on either side to protect the surface from scratches. Leave this film on while you drill and until you are ready to finally assemble the panel. It will protect the Plexiglas® from scratches during construction.

With all components cut to size, the next step is to drill the holes for the screws in the side bars, panel back, and plastic cover; and two holes for the take-off leads on the aluminum panel back.

For these particular panels I wanted to have four screws along each edge for a total of 16 screws to hold the structure together. More screws can be used if you wish. To drill the side holes, c-clamp the three components together, with the aluminum sheet on the bottom, the Plexiglas® cover on top of that, and the side bars on the very top.

Align all the pieces to each other and c-clamp all four edges so that nothing moves while drilling the holes. Then, drill the holes. Drilling the holes in this manner assures that the screw holes will line up as they should. You can drill each piece separately, but be sure that your measurements and drilling technique are very accurate.

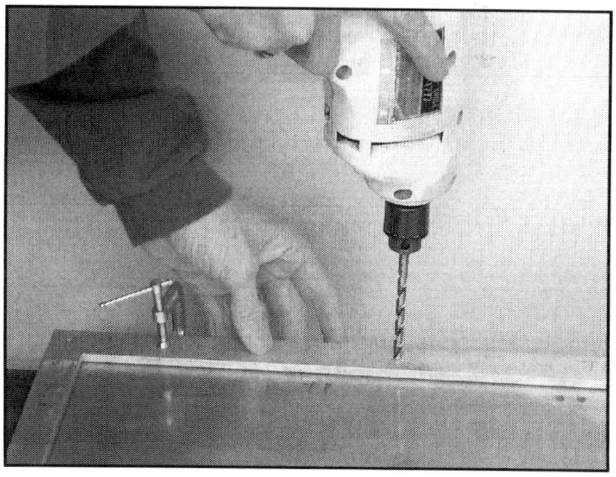

Clamp all the structural components together, then drill the holes for the screws.

Before you unclamp the pieces, mark each piece so that you will know exactly which piece goes where, when you are ready for final assembly. This is very important, as minor variations can cause assembly problems if the holes don't match exactly.

Power take-off box

After drilling the screw holes, measure, mark and drill the power take-off holes. One way to do this is to construct the power take-off box first. Any size project box from any electronic supplier or store can be used.

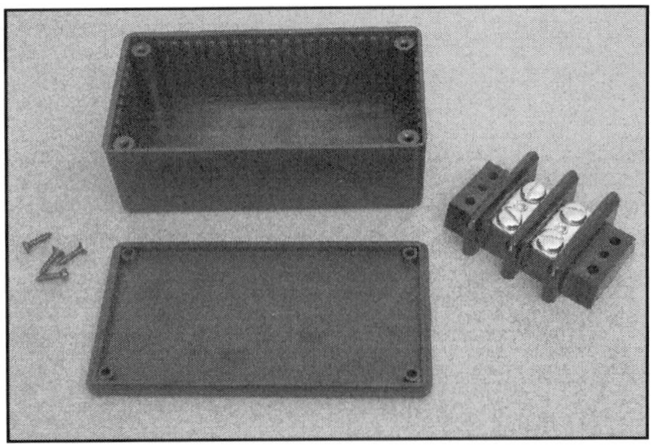

The power take-off box, parts (above) and assembled (below)

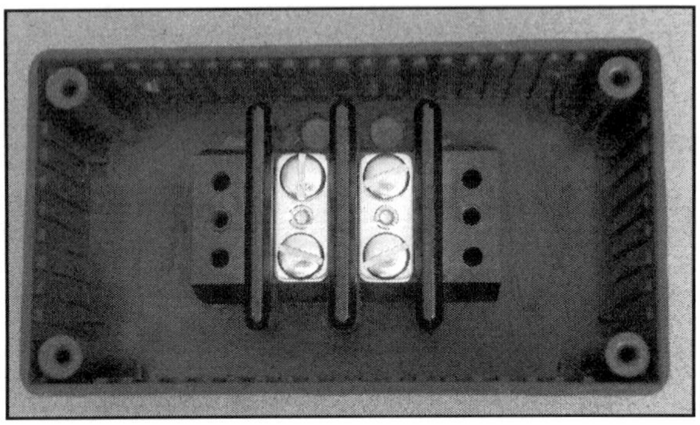

To prepare the box, first make sure the surfaces to be glued are flat. In other words, remove any mold marks from the outside back of the box, so that it will sit flat against the aluminum panel back. Sand the surface of the box to be glued to give added grip.

Then, remove any mold marks inside the box on the surface where the connector block will be glued. Place the connector block inside the box to determine the position for the take-off holes in the plastic box. Also, figure where the holes for the lead outlets to the electrolyzer will be. To choose an appropriate hole size, be sure to consider the width of the bus ribbon including shrink tube insulation. Outlet holes going to the electrolyzer should be sized for the wire used, plus additional space if you use rubber grommets in the holes. Rubber grommets can be slipped into holes to protect the wires from fraying and to partially seal the holes from moisture. Mark the take-off and the lead outlet holes and drill them.

Sand the inside surface of the box where the connector block will be glued, and also sand the back of the box where it will come into contact with the back panel surface. After sanding, clean the surfaces of all grit particles or other foreign material. Then, glue the connector block into the box with an instant glue or epoxy.

When the outlet box has dried, position it on the back of the panel where you want to put the take-off holes. Mark the take-off holes through the box onto the back of the panel and mark the outline of the box on the back of the panel. Drill the holes in the panel back for the power take-off ribbon.

Prepare the backing surface to lay screen

The next step is to sand the face of the panel back to lay the screen, which is mounted with silicone caulk. The sanding gives a rough surface that helps the silicone adhere. The bottoms of the bar stock (side bars) that will abut the panel surface should also be sanded for this reason.

After drilling and sanding, wash all metal parts down so that no metal powder or flakes are left on the surface. Tiny metal particles from drilling and sanding can short out the panel and cause problems later. After washing, be sure the parts are dry before applying the silicone.

Glue the side bars to the panel back

(See the photos on the next page.) Run several beads of silicone caulk along the face of the panel backing where the side bars will be placed. Insert the screws up through the panel back, then slide the bars on to the screws. Put the nuts on to hold the bars securely. This will hold the bars in place while the silicone is drying. Then, run your finger along the inside edge to smooth out any excess silicone

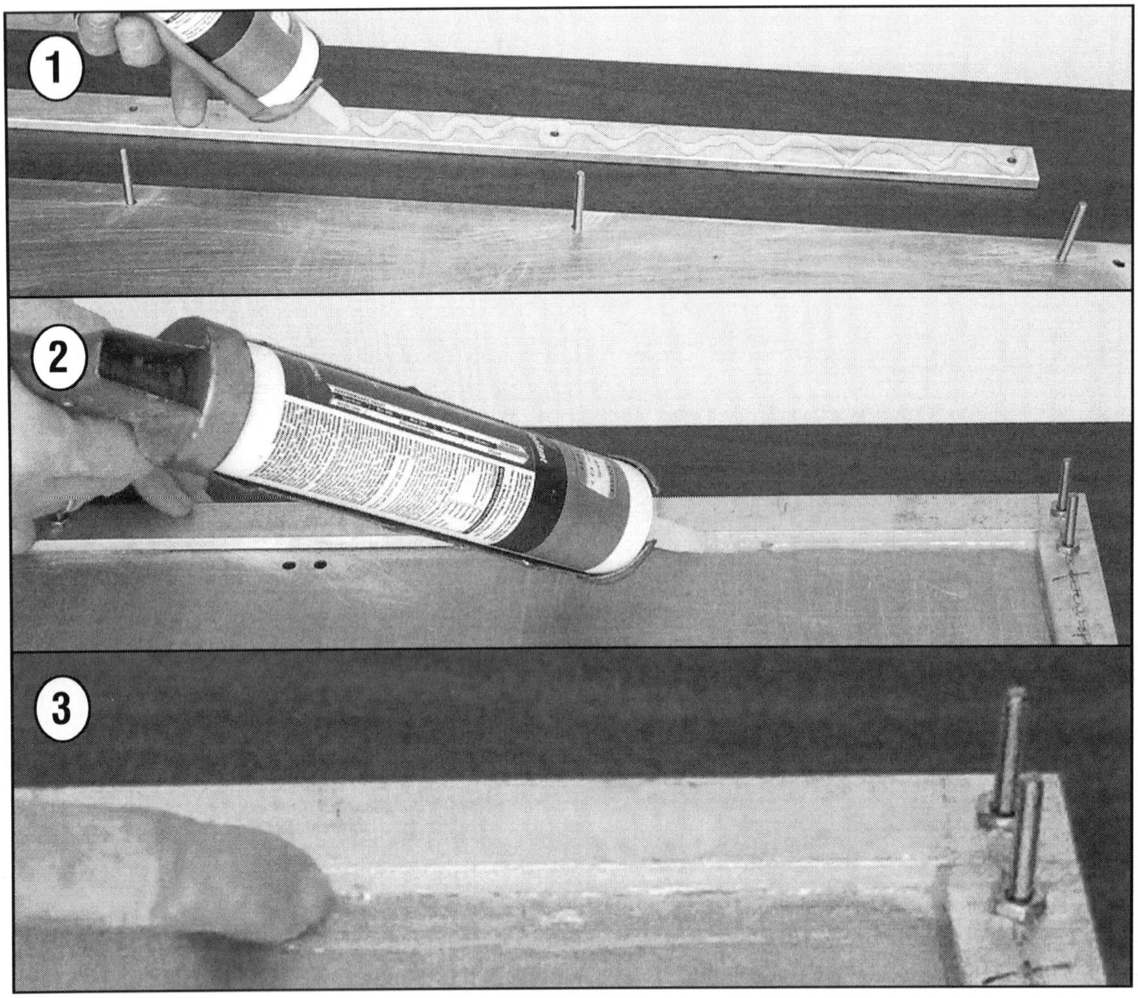

Glue the side bars to the panel backing

that is squeezed out from under the bars. Put another bead along the inside edge where the bar meets the back and smooth that out with a finger. This will create a moisture seal. Smooth the outside edge and then apply more silicone and rub along the edge with a finger to smooth it out. You can wear latex gloves for this. Give the silicone about 24 to 36 hours to dry, then take-off the nuts and remove the screws.

Apply the screen

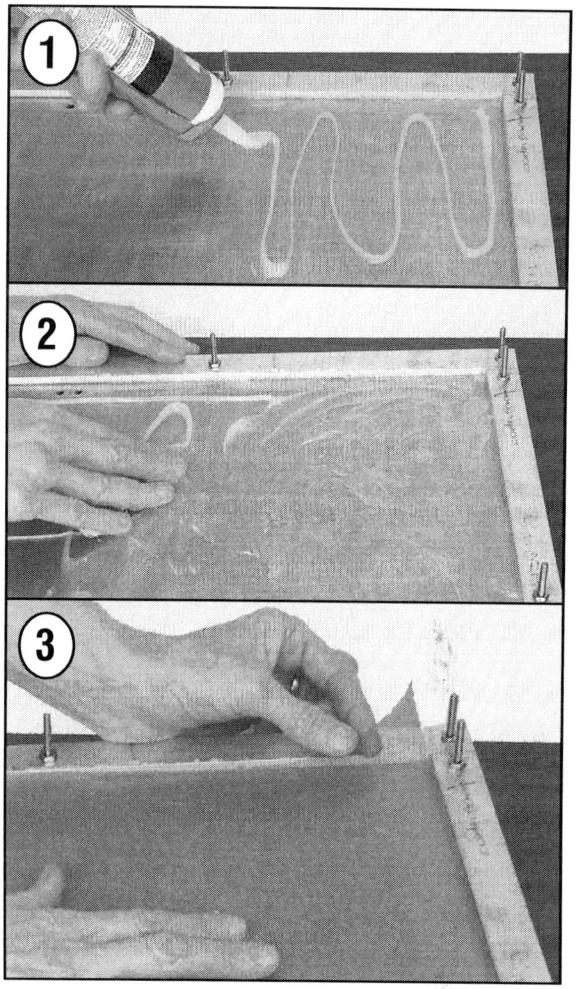

Apply a layer of silicone, spread it evenly, then lay the screen on it.

Cut a piece of fiberglass screen to fit within the frame/side bars. For these panels with overall dimensions of 28"x 30", I subtracted the width of the side bars (1" each) from the overall dimensions, so I cut the screen to 26"x 28". The screen insulates the PV cells from the metal back.

Cover the surface of the face of the panel backing with silicone caulk, in an even layer with no missed spots. The best way to do this is with your hands. Squirt the silicone from the caulk gun on to the panel back and them smooth it with your hands (see photos at right). Work quickly, because the layer of silicone is thin and will dry fast. Lay the screen in the silicone on the face of the panel backing. It should fit in perfectly. If not, trim any excess with a razor knife. Set it aside and let it dry for 24 to 36 hours.

Putting the cells into the frame

Peel/slide the cells into the frame. Use the fiberboard guide to gently slide the cells onto the frame, carefully aligning the cells to their final placement (see photo, next page).

Dab silicone along and under the bus ribbon of the connected strings (see below). Do this well, as this silicone is what holds the cells in place when the panel is upright. Be careful not to drop silicone on the cell surfaces.

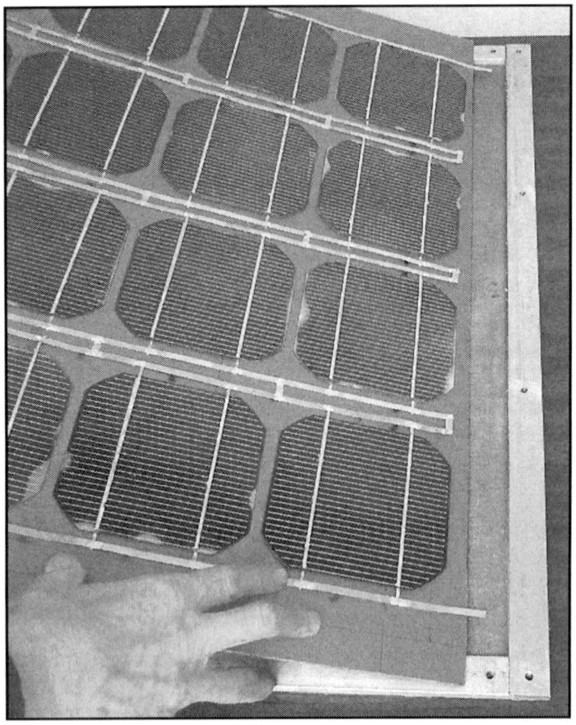

Slide the solar cells carefully onto the panel backing

Place 10 rubber spacers in place with silicone adhesive (see the illustration on page 25). These spacers hold the Plexiglas® away from the cell surfaces so that the cells do not break, and they support the Plexiglas® panel cover. They can also be made from heat resistant plastics. In placing the spacers, be sure that they do not cast a shadow onto the cells when the sun is at an angle to the panels.

Dab silicone under the bus wire to hold the strings of cells in place

Next, cut two holes in the screen for the power takeoff holes. Allow the silicone holding the cells to dry for 24 hours.

Solder the connection to the power take-off leads.

Take two pieces of bus ribbon (mine were about 16" long) and solder one to the left side at the top of the bus, and connect the other bus ribbon to the right side top. Put a piece of cardboard under the bus ribbons that you are soldering so that the heat from the soldering iron does not melt the nylon screen. (See above photo.)

Insert the power take-off leads through the take-off holes in the panel and notice where they touch the panel as they go through. Put shrink tubing on the take-off leads so that the leads are insulated from the metal backing of the panel. To do this, slip a piece of shrink tubing over both bus wire leads, hold the tubing in its final position, and apply heat to shrink the tubes.

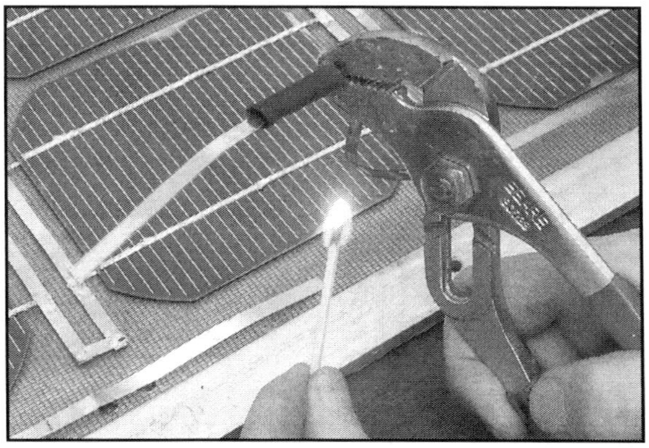

Use heat shrink tubing to insulate the leads.

Push the bus wire through the holes again and apply silicone under the take-off bus wire to hold it to the frame/screen surface. Then, dab a little silicone around the edge of the take-off holes, both front and back, for insulation. Bend the bus wire toward the panel back to hold it so that it doesn't flop around while it is drying. Let it dry for 24 to 36 hours before moving the panel to an upright position.

Attaching the power take-off box

With either epoxy or instant glue, cover the back of the box and the surface of the aluminum back that the box will adhere to. Push

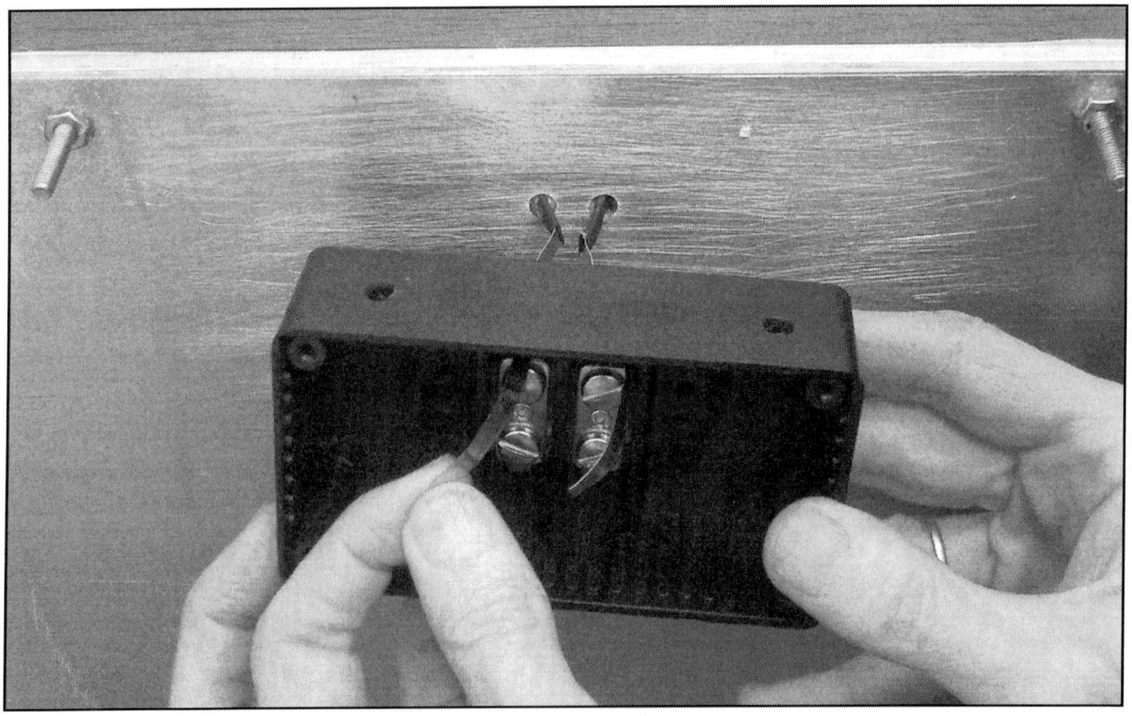

Pull the bus wire through the holes in the panel back and terminal box. (above and top).

the bus wire through the holes in the panel back and box. Position and press the box into place. If using epoxy, let it set before going on to the next operation.

Apply silicone all around the edges of the box that meet with the panel back. This will help to seal against moisture. Let it dry.

Attach the bus leads to the screw terminals. If the bus wire is wide, trim it so that it just slides into and under the screw heads in the terminal. Cut off the excess bus wire. Another option for wide bus wire is to drill a small hole in the bus wire for the screw in the terminal connection.

Placing the cover

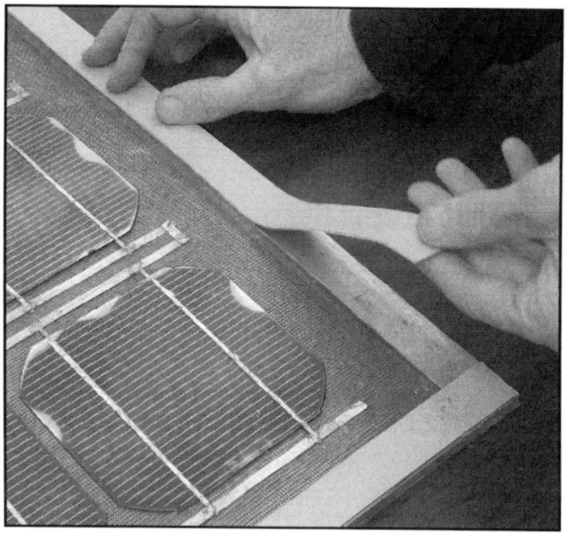

Silicone rubber strips are used to make a gasket between the side bars and the Plexiglas® cover

Adhesive backed 1/32" silicone rubber is used to make a seal between the side bars/frame and the Plexiglas®. This material comes with a paper backing, which makes it easier to handle, but once the paper backing is removed, the rubber shrinks a little in length. Leave the paper backing on initially to cut the rubber.

The 3'x 1' rubber sheet should be cut in strips the same width as the side bars. To figure the length to cut, two strips should be the height of your panel, plus an inch for shrinkage. The other two strips should be the width of your panel, minus the width of two of your side bars, then add an inch for shrinkage. This way, at the panel's

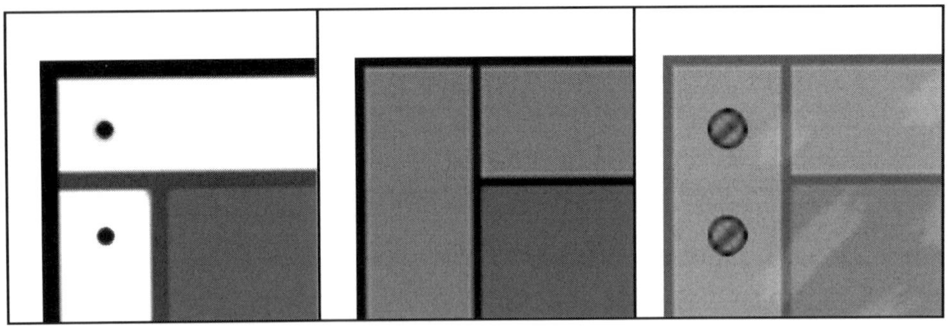

Three layers: aluminum side bars, silicone gaskets, Plexiglas®. Note that the gasket joint is staggered with the aluminum side bar joints.

corners the seams between the pieces of rubber are staggered with the seams between the side bars, which makes a better seal (see above). For each of my panels I cut two 31" strips and two 27" strips, all 1" wide.

Remove the paper backing from the rubber, and let it shrink. If you haven't already done so, remove the screws from the side bars. Lay the longer vertical strips of rubber on the side bars/frame, so that each strip covers the full vertical length of the panel, then trim off the excessrubber. Lay in the shorter horizontal strips, and trim them so that they fit snugly against the vertical strips. At the edges where the strips meet, apply silicone to improve sealing.

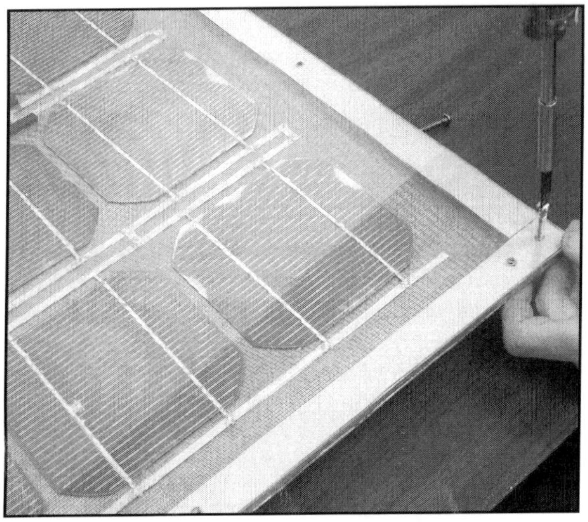

Installing the Plexiglas® cover

There are other ways to seal a PV panel. My book *Build Your Own Solar Panel* covers this in greater detail.

Remove the protective plastic from the Plexiglas®, and lay the

Plexiglas® over the panel, lining up the screw holes. Before inserting each connecting screw, dab a bit of silicone along the thread area and under the head of the screw. Also put a dab of silicone in the screw hole itself. Insert the screws down through the top. Before you put the nuts on to secure the panel, dab silicone where the screw comes out of the back of the panel, and then put the nut on. When the nut is tightened on each screw, dab some more silicone over the nut. This helps to make the panel water tight.

Finally, rub silicone along the edges of the panel to ensure a final water tight seal.

Finish wiring the panel

Connect the take-off wire (the wire that is going to the electrolyzer) through the holes in the junction box. If you use grommets, be sure to put those in first. Slip the wire through the box and either make a knot in each wire or use a wire hold of some sort to prevent the wire from being yanked from the terminal screw, from outside the box. Connect the red lead to the positive terminal and the black lead to the negative terminal.

Designing and setting up your system

Reducing circuit loss

Reducing circuit loss to a minimum is an important goal when designing a system. The three most important things to do to reduce circuit loss are:

1. Use the shortest wire run possible.
2. Use the largest diameter wire possible.
3. Reduce the use of diodes as much as possible, or use low voltage drop Schottky diodes.

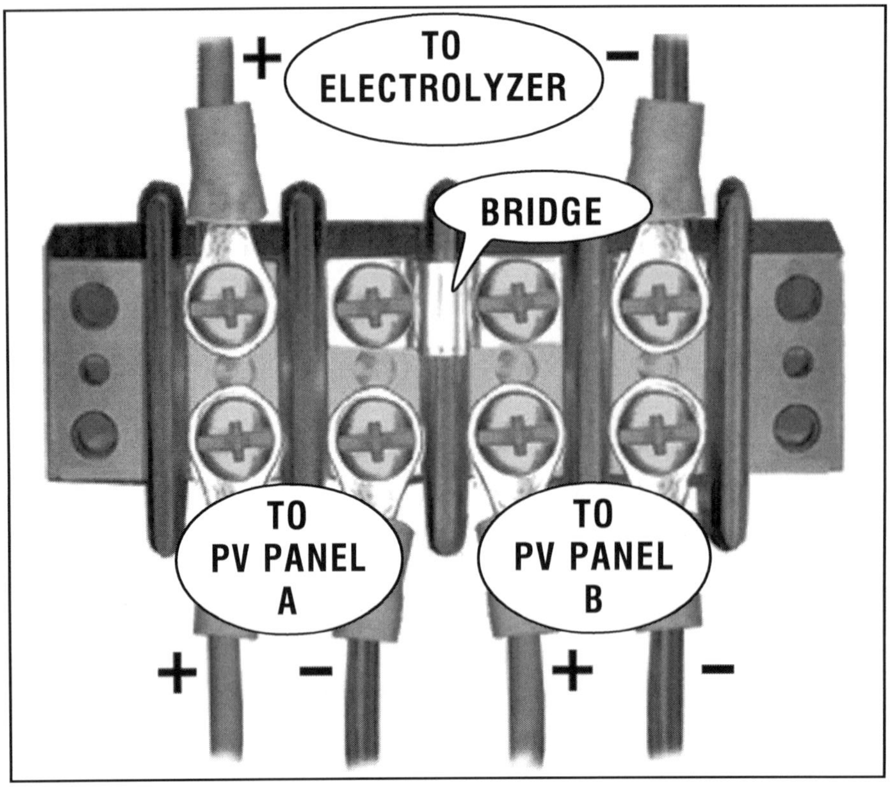

Connecting block configuration for two ESPMs connected in series.

The rating of the diode should always be more than the open circuit voltage and short circuit current produced by the individual panel or array used. Diodes also need aluminum heat sinks to dissipate the heat generated within the diode by the flow of current.

Connectors and switches should be of decent quality and all components should be sized correctly for current carrying capacity. The less resistance you introduce into a circuit, the less voltage drop you will experience.

Corrosion will create resistance in the circuit, so, it is important to house the components, switches, diodes, connectors, etc. in weatherproof housings if the components are to be out of doors for any length of time. A very slight oxidation on contacts can develop great resistance and reduce power output drastically. Connectors and

Configuration options for connecting PV panels to electrolyzers

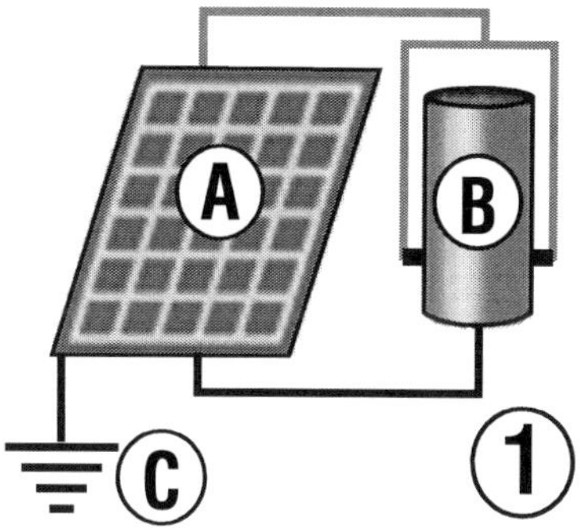

A	Solar Panel
B	Electrolyzer
C	Ground for panel frame
D	Diode
E	Toggle Switch
F	Lightning Arrestor
G	Diversion Controller
H	Battery
I	Fuse

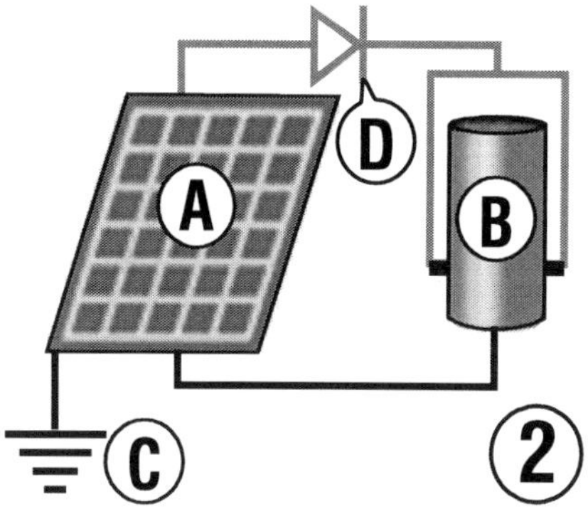

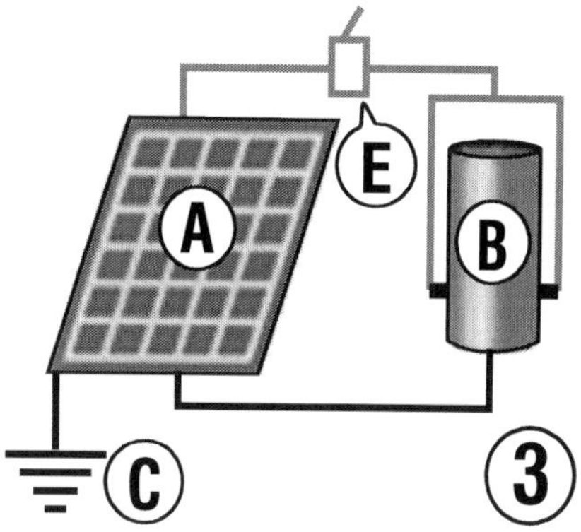

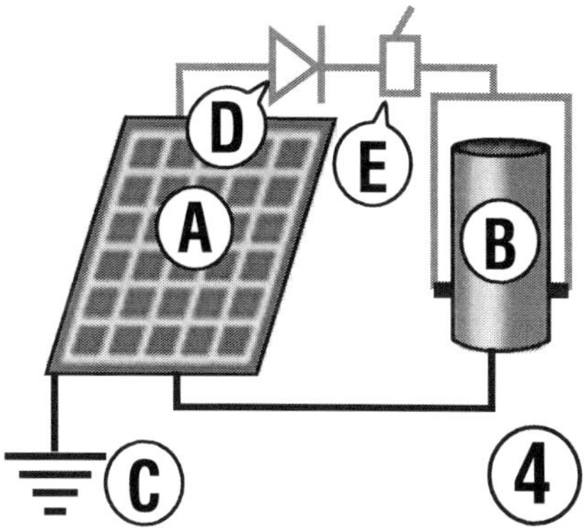

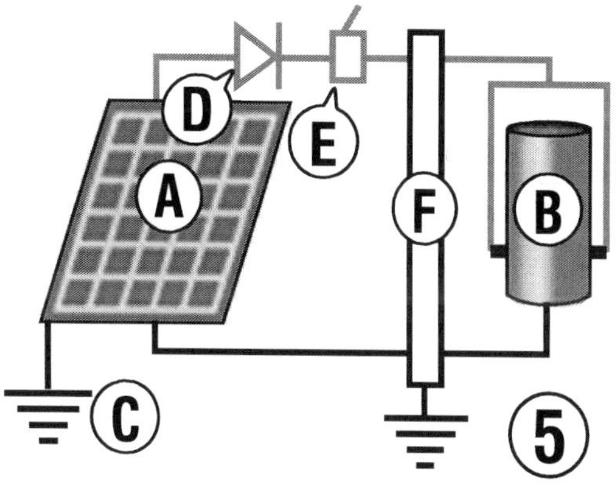

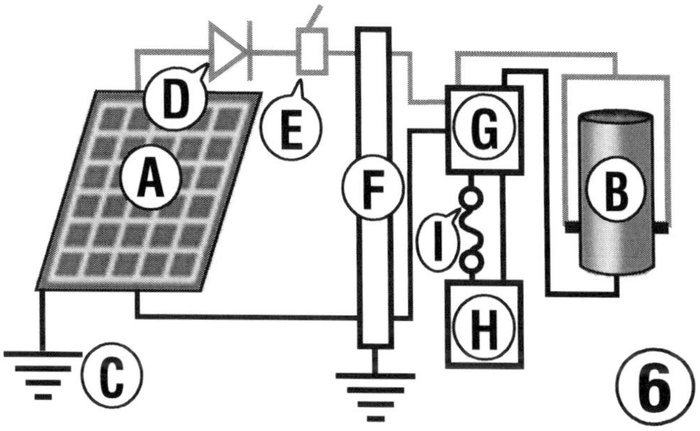

switches need to be protected from moisture and rain to prevent short circuits, so make sure the housing for them is wet proof. Most electronics and electrical suppliers have boxes with rubber seals that are made to be weatherproof. You can also seal boxes with silicone caulk. Connectors can be covered with liquid electrical tape or a dip type rubber compound to insulate them.

Grounding

ESPM panels should be grounded for long term service to protect against lightning strikes and electrostatic charge buildup. Even distant lightning strikes can induce EMF in the panels and cause problems. Wind, pollen, dust, and snow blowing across the panels can also cause electrostatic charge buildup. Grounding braid can be connected to the panels on one of the long screws that jut out on the back of the panel, with the other end of the braid securely attached to a ground rod. To add circuit grounding, use a lightening arrestor for added protection.

Diodes

With the simple system described in this book, no blocking diode is needed because the "dark current" (night time reverse flow) is negligible from this electrolyzer and will not harm the panels. For a large bank of electrolyzers, consider putting in a blocking diode. Blocking diodes are used in BSPM systems primarily to prevent back flow of current from the battery into the panel at night, which depletes the batteries. However, there is some question about the usefulness of blocking diodes for less than 24 volt BSPM systems because charging current is lost through the diode during battery charging anyway.

A solar hydrogen system that does not have a hybrid storage system (both batteries and gas storage) or doesn't use BSPMs exclusively, will have no loss of stored energy back into the PV panels. Diodes can be used, but, for this system they are not necessary.

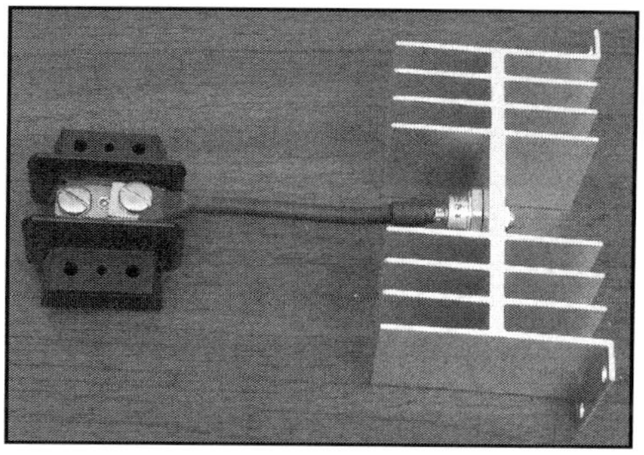

Blocking diode
with heat sink

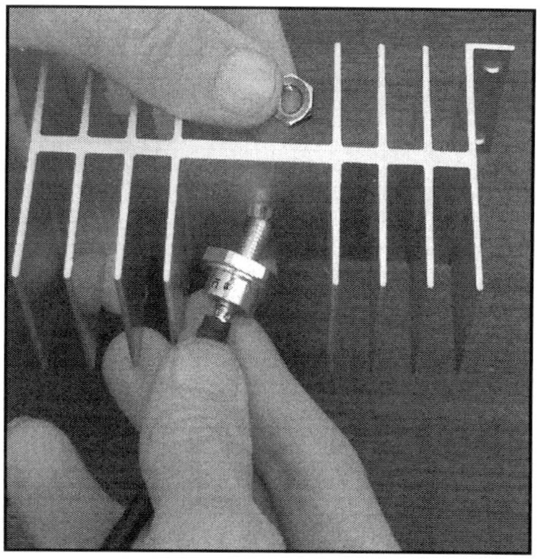

Switches

An on/off switch can be useful. It can be as simple as a weatherproofed toggle switch rated above the amperage of the panel or array. Such a switch in a solar hydrogen system should be placed close to the panel and not near the electrolyzer unless it is an explosive rated switch. A simple connector box which contains the contacts and/or diodes can also house the switch.

A low cost wireless switch can be installed in the box to turn the electrolyzer on or off at a distance. A wireless float switch could be used to turn off the electrolyzer when the gas tanks are filled. The details of such a switch will depend on the use and/or storage methods for the gases involved.

Data loggers

Data loggers can be integrated into the system to record volts and amps. The data can be downloaded to a computer from the field, or be directly connected, either hardwired or wireless, to transmit data to the computer. This will give a very accurate record of the system's performance on a minute to minute basis, and give you the capability to correlate panel output to gas production, daily insolation, etc.

Simple voltmeters and ammeters can be added to the circuit for quick visual indicators.

Fuses, connectors and cables

If you are interested in a hybrid storage system with a battery, you can use a blocking diode and you should definitely fuse the battery, or use MCBs (miniature circuit breakers) to prevent possible problems from a short.

Battery connections should be very secure. Do not use alligator clips as they have a tendency to be easily knocked off and spark. Always use weather proof boxes for connections and switches. If you are going to run cable underground and or have it exposed to sunlight outdoors, make sure it is rated for that purpose, that is, type TC or UF. This cable can be wired to the power box where you have the switch.

For panel to panel connections, a #12 or #10 red and black zip cord is fine, and, depending on your setting, may be all that is needed for your connections. If you do not have underground runs or extensive (more than 10′) outside runs, or you do not expect to keep the system up for extensive periods of time, then the zip cord is probably all you need. The integrity of your wiring should be checked regularly. Outdoor wiring can be coated for UV protection. Ultimately, refer to any codes for your local area for the final word on wiring requirements.

Positioning solar panels

To maximize the output of a solar array, angle the panels according to your latitude and to the season. To find your latitude, look at a paper map or atlas, or access the map area of the US Geological Survey website. There are also many other sites on the internet that give this information for wherever you live in the world. On the USGS site, for instance, all you have to do is put in the name of your town, state and country.

Where I live in Vermont in the northern hemisphere, during winter the sun is low on the horizon as it transits east to west. During the summer the sun is much higher and transits overhead. In the spring and the fall it transits in between the lowest winter point and the highest summer point. My latitude is 44 degrees and 35 minutes north. The optimal angle for any latitude is the latitude angle itself. At my location, that's 44° 35' from the horizontal. I can leave this angle the same for the whole year, but to take best advantage of the sun's changing position, I adjust the angle four times a year, halfway between each solstice and equinox, which is about May 5, August 5, November 5 and February 5. The routine is, on August 5 and February 5 the tilt is the same as your latitude, 44° 35' for me. November 5, add fifteen degrees tilt to your latitude for the winter solstice, for me 44° 35' plus 15° which equals 59° 35'. May 5, subtract 15° from your latitude for the summer solstice, for me, 44° 35' minus 15°, which equals 29° 35'. In practice I round off everything and tilt my panels to 45° on February 5 and August 5, 60° on November 5, and 30° on May 5.

Figuring the optimal angle for solar panels in the northern hemisphere.

The panels also have to be oriented facing true south. The fastest way to calculate true south is go to the Natural Resources Canada website and use their magnetic declination calculator to obtain what is called the magnetic declination for your longitude and latitude. The magnetic declination is the difference in degrees, minutes and seconds between magnetic north and true north. To use the magnetic declination figure to orient PV panels, use a compass to find magnetic north, then add or subtract the magnetic declination according to your location. Orienting the panel to true south and making adjustments for seasonal variations will enhance hydrogen production.

Solar trackers

To increase production further, you can build or buy a solar tracker. A solar tracker will follow the sun in its seasonal declination, as well as track its daily east to west movement.

Trackers are a good addition to any solar hydrogen system. For a PV system that has only battery storage, generally trackers do not improve efficiency for much of the summer season. This is because battery banks typically get fully charged in just a portion of a summer's day. But in a hydrogen system, all the energy the PV panels can convert from sunlight to electricity can be used.

There are several types of trackers available including passive solar and solar powered motor driven. They are not at all necessary for a system, but they are an option to increase hydrogen production, if you want to experiment with them. They are especially useful in very sunny climates. Their value is probably not as great in cloudy climates, since the sunlight is more diffuse. Some tracking systems take some outside energy to operate, and thus must have their own PV power supply. Trackers also add to the complexity of a system.

Remember, anything mechanical requires more care and maintenance, especially in harsh weather conditions and climates.

Mounting PV panels

PV panels can be mounted in many different ways. They can be secured to racks that stand on poles, or be ground mounted, or put on roofs, for instance. I would be cautious about any roof installation. Panels need to be cleaned regularly, and they must be kept free of snow and ice. If they are not easily accessible, this will not happen. I suggest short pole or ground mounts, and if you live in snow country be sure the panels are high enough off the ground to keep the panels out of the highest snow cover – but low enough to remove the snow easily.

Panels should never be put in an area that is going to be shaded – not even a little bit of shade. If a portion of a cell gets blocked from sunlight, it can take down most of the output of the entire panel and can cause other more severe problems. Diodes used in a bypass fashion can alleviate these problems, but they will add to system loss and should be avoided whenever possible.

Make sure your PV panels are secure in high winds. Panels make great sails and must be sturdily fastened to whatever mounting system you use.

Always leave plenty of air space at the back of the panels so that air can circulate and cool them. They get very hot in use, and heat reduces their efficiency and can stress the materials, especially your soldered connections, so do whatever you can to keep the panels as cool as possible.

If you can, avoid locating panels near dusty roads. Also, clean panel surfaces at least once a week with plain water. Barely visible dust and pollen can collect quite rapidly and affect the output of the panels.

My book *Solar II* has more detail about designing and setting up PV systems - mounting and wiring PV panels and other system components, etc., which you may find helpful.

Final design considerations

The final design of your panels should be based on your intended use. Panels can be constructed in a number of different ways, and with variety of materials.

For this particular project I did not build the ultimate panel, as I wanted to use up materials I had on hand, such as the aluminum backing material. This use limited the panel output because it limited the number of cells that could fit within the frame. If the materials on hand had not been an issue, I would have made two panels with an output around 2.5 to 3 volts, or one panel with a 4 to 6 volt output. However, my yankee sensibilities ruled and I figured the smaller size would be adequate for a prototype to run the electrolyzer.

The main consideration, if you build the electrolyzer in this book, is to try to have 4 volts at about 20 amps for each electrolyzer. You can definitely work with less, but due to circuit losses, varying sun intensity and so forth, this is a good minimum to aim for.

Consider less than optimal conditions in your design. If you are looking at long wire runs and diode use, plan this into the whole system.

If you wish to purchase ESPMs rather than build your own, be ready for some sticker shock. While it is inexpensive to build these panels, it is not inexpensive to buy them. The reason is that the photovoltaic industry has not yet caught up with the idea of the coming hydrogen economy. This is no surprise, since many PV companies are owned by large oil companies, and for a variety of reasons they are not interested in promoting small scale hydrogen production.

The PV industry is geared for BSPMs, which means all of their tooling and assembly line processes are set up to efficiently and economically produce battery charging solar panels, but when the panel parameters are changed, it becomes a special order and thus very expensive, even though the materials used are identical. Hopefully this will change in the near future. If you are involved in the photovoltaics industry, consider this present void an economic opportunity, and a chance to open up the industry to these possibilities. It would not take much research and development to bring some good products for hydrogen production to market.

Electrolyzers

In a solar hydrogen system, the electrolyzer is the component which changes electrical energy from the photovoltaic panels into hydrogen gas.

There are many types of electrolyzers: high temperature, high pressure, low temperature and low pressure, and liquid electrolyte and solid electrolyte forms. For solar hydrogen production, low to medium pressure, low temperature liquid electrolyte electrolyzers are preferred. When compared to the cost of high temperature, high pressure systems and/or solid electrolyte systems, they are inexpensive to make, purchase and maintain.

Solid electrolyte PEM vs. alkaline

Solid electrolyte PEM (proton exchange membrane) electrolyzers can be used in systems to avoid use of caustics as an added safety factor; and where no one is available to frequently monitor a fluid electrolyte system. PEM electrolyzers are much more expensive, and do not have the track record that alkaline electrolyzers have in use. Although they are reportedly almost trouble free during use, they do pose problems in terms of cost of replacement parts when they become inoperable. Failures in PEM electrolyzers are usually membrane blow-outs or catalyst degeneration. Both problems are costly to service with replacement parts.

Alkaline electrolyzers, on the other hand, use a very inexpensive electrolyte and low cost electrodes such as nickel which are easy to obtain and replace. Eventually, as lower cost alternatives to presently expensive ionomers and catalysts become available, PEM electrolyzers will come to the forefront and probably be more widely used than alkaline.

Cost was an important consideration for us in the development of our experimental solar hydrogen system, so a PEM electrolyzer was discounted in favor of a low pressure alkaline tank electrolyzer. High pressure and high temperature electrolyzers were also evaluated. These were rejected due to higher amounts of energy consumed and the need for more intensive monitoring of the system.

Electrolyzer basics

An alkaline electrolyzer, as used in this system, is a simple electrochemical device that disassociates water into its constituent molecules, oxygen and hydrogen. This is accomplished by the application of very low voltage and high amperage DC (direct current) electricity in an alkaline electrolyte solution consisting of potassium hydroxide (KOH) and distilled water.

The term electrolyze means "to loosen with electricity." The term is derived from the ancient Greek word elektron which means "amber." When amber is rubbed, electricity is produced. The word lusis means "to loosen." Thus, together they mean "to loosen by electricity."

Although an oversimplification, DC electricity flows in one direction only, which is what is needed for the electrolytic process. AC (alternating current) electricity, such as used in our homes, flows in both directions at a specific frequency and is thus not suitable for use in the process of electrolysis unless it is rectified.

DC (direct current) electricity, which powers the electrolyzer, can come from a variety of renewable sources, such as wind generators, photovoltaic panels and small hydro and geothermal systems. Batteries are a common non-renewable source of DC current, as well as AC power supplies with rectifiers that change the AC to DC.

An electrolyzer consists of two electrodes, usually made of pure nickel, a nickel iron alloy, stainless steel, monel, or Raney nickel. One electrode is connected to the positive source of the DC power supply and the other electrode is connected to the negative source of the DC power supply.

The electrodes are immersed in a potassium hydroxide (KOH) solution in a tank. The tank has collector tubes to carry off the generated gases. Hydrogen is produced at the negative electrode and oxygen is produced at the positive electrode.

There are many factors which determine the amount of hydrogen and oxygen generated in the electrolyzer. The most important considerations are:

1. Percentage of KOH to water in the electrolyte solution.
2. Surface area of the electrodes.
3. Physical distance of the electrodes from each other.
4. Amount of DC current (amperage) applied

A higher ratio of KOH to water increases conductivity up to 29.4% of KOH in solution. After this point, the resistance increases and there is no point in adding more KOH.

The greater the surface area of the electrode, the greater the gas production will be. However, the greater the surface area, the greater the amount of current that is needed to realize the full potential of the added surface area. Hydrogen production, as it relates to current density, is calculated by dividing the amount of current by the electrode area. This is expressed as amps per area. Each electrolyzer has its own operating parameters for its most efficient operation.

Porous alloy electrodes

Porous alloy electrodes of Raney nickel are often used in electrolyzers. Raney nickel is produced by first making an alloy composed of 50% aluminum and 50% nickel. This composite is then treated with potassium hydroxide, which eats away the aluminum and leaves a porous nickel sponge material, known as Raney nickel after Murray Raney, the inventor of the process. Electrolyzer electrodes of this material have a large surface area due to their porosity and will produce more gas with smaller electrodes, compared to electrodes made of materials with less surface area, such as sheet or screen. Although Raney nickel is preferred, it is more expensive than either sheet or screen. The porous texture that creates a larger surface area also acts as a filter for small particles. The sediment which forms reduces the active surface area over time, inhibiting gas formation and thus efficiency.

Prepared surface flat plate electrodes

The surface area of flat plate electrodes can be augmented by sanding or sandblasting the plate, thus improving gas production. Simple sanded flat plates have less surface area than Raney electrodes would provide, but they have more surface area than a simple flat plate. Besides being economical, they can be reconstituted quite easily if needed, by resanding, and so would last longer in operation. Use a very course grit sand paper and apply to the electrode surface in a multidirectional pattern to get as much extra surface area as possible.

If you want to experiment with these kinds of electrodes, a nickel-iron alloy foil that is 80% nickel can be purchased from McMaster-Carr. This particular foil makes an excellent electrode and is relatively inexpensive.

Mesh electrodes

Another alternative for electrodes is fine mesh screen. This provides more surface area than the flat plate and is not as easily clogged up as is the case with nickel sponge. Either 316 stainless steel, monel or nickel screen can be used. Nickel screen is expensive, so the 316 or monel is a more cost effective choice. The best choice is the monel, an alloy of 65% nickel, 33% copper and 2% iron. This alloy has excellent corrosion resistance in alkaline solutions and will last a long time.

Electrode spacing

The closer the electrodes are to each other, the more copious the gas production. However, the closer the electrodes, the more risk there is of mixing gases, even if membrane separators are used.

Some commercial electrolyzers sacrifice gas purity for electrical efficiency with closely spaced electrodes, but they require expensive purification equipment at the end of the process, which negates any cost efficiency at the production point. The more cost effective solution for purer gas production is to err on the side of separating the electrodes a bit more, and sacrifice a certain amount of electrical efficiency.

The more current, the more gas will be produced, within reasonable limits. All electrical systems have voltage and current parameters that, if exceeded, will deteriorate production or destroy the equipment. The current carrying capacity of the electrodes is a factor here.

A minimum input of about 1.29 to 1.49 volts is needed to initiate the process of separating the hydrogen and oxygen that constitute the water molecule. Voltage applied to the electrodes is usually in the range of 2.5 volts to 6 volts. The electrolyzer in operation will usually draw between 1.7 to 4 volts.

The electrolyte

Potassium hydroxide (also known as caustic potash) is a strong electrolyte. This means that it is essentially 100% ionized in solution and thus is a good conductor of electricity. When the positive and negative poles of the electrolyzer are connected to the power source, hydrogen ions combine with electrons at the negative electrode to form hydrogen, and hydroxy ions give up electrons at the positive electrode, releasing oxygen. Twice as much hydrogen is generated as oxygen, since the water molecule contains two hydrogen atoms for every oxygen atom.

Other electrolytes, such as sodium hydroxide, can be used, but they are not as conductive as KOH. Acids such as sulfuric acid can used as electrolytes, but this is more corrosive to the electrodes, and the added wear and tear on various components does not justify its use.

Essentially KOH is the best choice for alkaline electrolyzers.

KOH can be purchased, or made if you have a source of hardwood ashes. The making of lye used to be a common chore in most households. If you are interested in making your own KOH my book *Build Your Own Fuel Cells* contains complete illustrated instructions. If you buy KOH from a chemical supply house you will find the following table helpful.

If you make your own KOH, use this table to determine the specific gravity of the solution when you take a hydrometer reading. This will indicate whether to boil the solution down more to strengthen it, or add more distilled water to weaken it.

I make my own KOH from wood ash and do not bother to boil it

Potassium hydroxide (KOH) solution strength

Specific Gravity	Percent KOH	Lbs. per US Gallon	Specific Gravity	Percent KOH	Lbs. per US Gallon
1.0083	1	0.0841	1.1493	16	1.535
1.0175	2	0.1698	1.159	17	1.644
1.0267	3	0.257	1.1688	18	1.756
1.0359	4	0.3458	1.1786	19	1.869
1.0452	5	0.4361	1.1884	20	1.983
1.0544	6	0.528	1.1984	21	2.1
1.0637	7	0.6214	1.2083	22	2.218
1.073	8	0.7164	1.2184	23	2.339
1.0824	9	0.813	1.2285	24	2.461
1.0918	10	0.9111	1.2387	25	2.584
1.1013	11	1.011	1.2489	26	2.71
1.1108	12	1.112	1.2592	27	2.837
1.1203	13	1.215	1.2695	28	2.966
1.1299	14	1.32	1.28	29	3.098
1.1396	15	1.427	1.2905	30	3.231

down. I simply refill the reservoir with more KOH solution rather than distilled water. The KOH, when I first make it, comes out at about a 12% solution. In the electrolysis process, the solution becomes stronger as the water disassociates and is used up, leaving the KOH behind. Once the solution in the electrolyzer reservoir is at the right specific gravity, all that is necessary is to add distilled water as the water in the electrolyte disassociates and is used up.

If you start with the exact specific gravity that you want when you first fill the reservoir, then afterward, simply replenish the reservoir with distilled water to the same level of your first fill. KOH is lost over time, but it is mostly the water that is used up in the process. Also, be sure to use distilled water only. Well, tap, and spring water contain far too many unknowns (minerals, organic particles, etc.) that will cause problems with the electrolyzing process and gum up the electrodes.

Never mix dry purchased KOH into the water in the reservoir – the process generates too much heat. Always mix KOH into the water by putting a little bit in at a time, very slowly. Do not mix by pouring water onto the KOH. Use only plastic or stainless steel buckets or containers. Do not use aluminum pots or utensils for mixing or holding KOH solution!!! And, always let the solution cool down before refilling the reservoir.

For best conductivity, the solution should be about 29.4%, which means it would have a specific gravity of about 1.28. This would require 3.231 pounds of KOH for one gallon of distilled water.

This particular specific gravity is not necessary, and a milder, less conductive solution of about 12% will work. For optimal performance however, you will want to work with a 29.4% solution. To check the specific gravity of a solution, use a hydrometer such as those designed for testing the specific gravity of battery electrolyte. Hydrometers can be purchased at any auto parts or hardware store.

Safety

If you work with KOH, never forget that it is extremely caustic and can cause severe burns and blindness if not handled properly. Please do not ignore these warnings. Make sure you have eye protection such as safety glasses or a safety face shield. Either can be purchased inexpensively at most hardware stores. Always have all your skin completely covered with protective clothing and rubber gloves when working with and around KOH; and study and follow all MSDS recommendations. (See page 199). Always diligently follow all safety precautions when handling KOH, and keep it out of reach of children and pets.

In this particular system, gas pressure is supplied by a hydrostatic column. This is simply a filler tube with a reservoir on top at a height of five to eight feet, depending on how much pressure you want in the system. Because the reservoir is at eye level or above, it must be very well secured. You absolutely do not want nasty KOH raining down on you during your experiments. Hose clamps should be used to secure the tubing to the reservoir so that hoses cannot slip off and squirt KOH all over the place.

Water for the electrolyzer

The electrolyzer that you use in your system needs distilled water. Distilled water is for the most part contaminant free. Municipal tap water usually contains chemical additives, and well water contains dissolved minerals which, over time will encrust the electrodes and impede the action of the electrolyzer, so neither of these is suitable for electrolyzers. Rain water or melted snow are second choices to distilled water, although there are still some contaminants in these as well. The water for an electrolyzer should be distilled, whether you buy it, or make it yourself.

Solar water distillation

One of the best and most economical ways to distill water is with a solar water distiller. They are easy to make and, if you have the space, can supply some or all of the distilled water needed for the electrolyzer system.

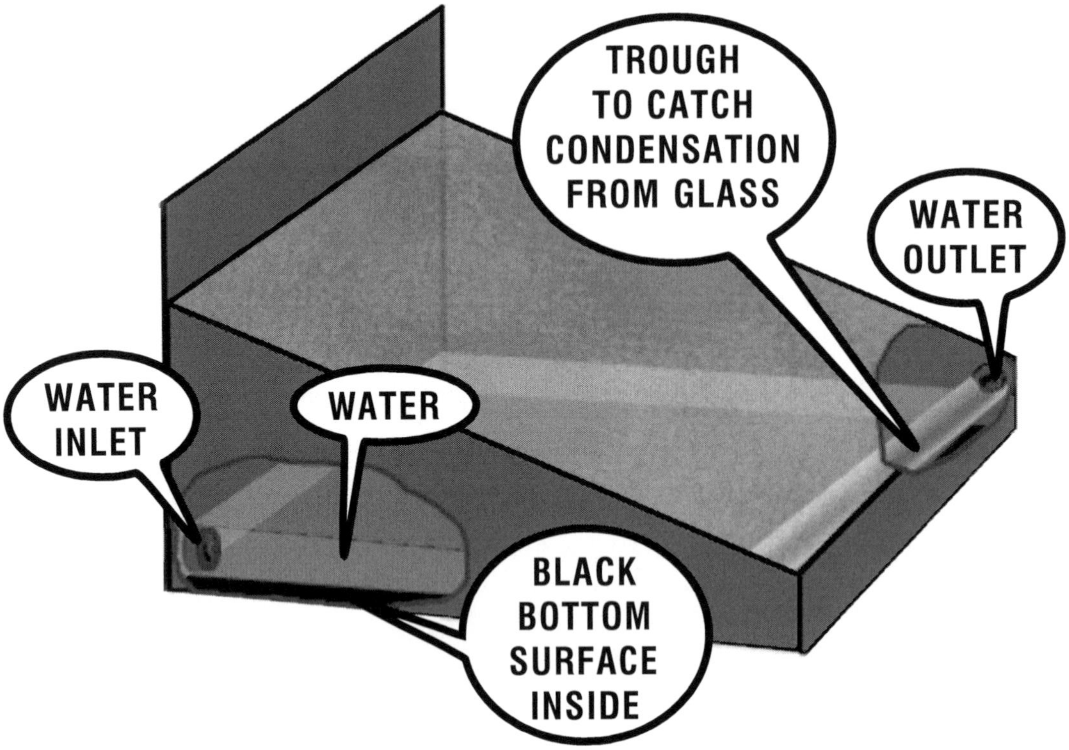

Simple solar water distiller

The illustration above shows a very basic solar water distiller. This is a very simple device that evaporates the water, leaving any mineral constituents behind. The water condenses on the glass and then drops into the trough and then out through the tube into a collecting bottle.

The glass cover can be a used patio door or any other large-sized piece of glass. The box is easy to construct, and can be made of wood or metal. The water trough inside the distiller can also be made of wood, metal, or heat resistant plastic. Tubing should be silicone, and reservoirs and collection bottles, etc. can be plastic or glass. The inside of the box has to be lined with a UV (ultraviolet) resistant and non-contaminating material, and needs to be black in color so that it absorbs heat and is leak proof. Black silicone caulk can be used to cover the entire inner surface, or some type of rubber or plastic liner that will not degrade can be used. The glass top should lie flush on the box with a tight fit to keep the heat in so that water vapor forms on the glass.

Volunteers In Technical Assistance (VITA) has a booklet by W. R. Breslin titled *Solar Still* which has instructions for building a solar heated distillation system.

The P41 electrolyzer

The P41 is a simple electrolyzer specifically designed to provide maximum output with the smallest footprint possible with off the shelf components. As with every part of this project, the important factors were ready availability of components at low cost, and ease of working with the components, so that anyone with average skills could build a device that would be comparable with commercial equipment.

For this electrolyzer, the primary goal was to produce enough hydrogen to run a real time stationary fuel cell system within the smallest possible space without the gases intermingling.

Design for renewable energy power sources

An important design consideration was to maximize the efficient use of energy input from renewable energy power sources. Most commercial electrolyzers are designed to run off utility power grids with a rectified DC source. The electrode materials used in these electrolyzers reflect that particular type and quality of power source.

Making intermittent power efficient

DC power from renewable sources such as photovoltaic panels is delivered intermittently. Depending upon where you are, there are very sunny clear days, but there are usually more days when the sun is behind the clouds for five minutes then bursts out in full sun only to be behind another cloud in another five minutes. This varying amount of sunlight will cause the disassociation of hydrogen and oxygen from water to proceed at a faster or slower pace according to the unsteady flow of DC power from the PV panels. This causes sharp spikes and troughs in gas production. This is not a problem per se, but it led me to think of nano-nuances in regard to renewable energy delivery systems.

A minimum voltage of about 1.23 V at about 77°F will disassociate water into hydrogen and oxygen, but this process absorbs heat. At a voltage of about 1.49 at around 77°F, no heat is absorbed. Higher voltages than 1.49 release heat during the disassociation process. Elevated operating temperatures increase the efficiency of the electrolytic process because less electrical input is needed.

Electrode materials

Experiments with a variety of materials led to the conclusion that monel is a perfect choice for electrodes in an electrolyzer designed for intermittent power supplies. It has an electrical resistance of about 42 micro ohms-centimeters at 20°C. This is much higher than, for instance, nickel, which has a resistance of 11. With higher resistance, more heat is generated as the current passes through the electrodes and electrolyte solution.

Thermal flywheel

For conventional electrolyzer design, monel would not be considered to be as efficient as other materials – the heat generated would be considered wasted energy. However, for intermittent renewable energy systems, monel is a good choice because thermal input dissipates at a much slower rate than electrical input. Thus, the thermal energy is still feeding the process when, for instance, the sun is momentarily obscured by a cloud. A "thermal flywheel" is created that retains the energy gained from the photons as heat, which lowers the threshold for the electrolytic reaction in terms of how much voltage and current is needed. When the sun is behind a cloud, less electricity is generated, but the heat retained in the electrolyzer requires less electricity to continue the reaction at a more active level than without the additional heat.

Supercapacitors

Another approach to consistent power delivery from renewable sources is to consider the use of supercapacitors placed between the renewable power source and the electrolyzer. Supercapacitors are basically a cross between capacitor and battery technology. They use electrodes, and a liquid or organic electrolyte, but they store energy by static charge rather than by electrochemical means. They can be cycled millions of times, and have a recharge time of seconds. Supercapacitors can also enhance peak load performance on the fuel cell end. For more on this, see my book *Solar Supercapacitor Applications*.

Micro and macro electrode surface considerations

The next design consideration was surface area. The greater the surface area of an electrode, the greater the gas production will be for a given space.

Raney metal surfaces

At the present time, Raney metal structures expose the greatest amount of surface per area. (Nano-tube structures will soon surpass Raney efficiencies, but right now these nano-tube electrodes are quite expensive.) However, as stated earlier, Raney structures have the nasty habit of acting as extremely fine porous filters that gum up. This would reduce the efficiency of an electrolyzer incrementally over time. Also, the micro porous Raney surfaces seem to retain gas for longer periods of time in pockets, which blocks the reaction and renders a portion of the electrode surface useless. This creates a higher active threshold, which is acceptable for industrial systems that run off consistently supplied grid power, but is not particularly suitable for intermittent renewable power supplies. Another disadvantage is that Raney surfaces are more expensive than other alternatives. They are mainly used in industries where cost of electrode replacement is not a concern. They make excellent electrode surfaces, but keep in mind the degeneration that can occur in a not too perfect environment, and that replacement of electrodes should be considered in your design.

Raney electrodes did not have the profile we were looking for. We needed a very inexpensive material that was rugged and would last almost forever. 200 monel mesh seemed to be the right answer. It is extremely resistant to corrosive environments, provides a larger surface area than flat or surface sanded plates, and it does not retain gas bubbles for long periods of time so that the same surface could react more frequently

Electrode shape

After considering the micro surface, the next task was to design the macro surface, which is the shape of the electrodes themselves.

After much consideration, we came up with a star-pleated design for the negative electrode inside a plain cylinder positive electrode. In this way, we could match the surface areas of the two electrodes in a small amount of space. There are detailed illustrations of the electrode design later in this chapter.

We wanted to have as much macro surface as possible, but at the same time did not want to overdo it. We expected better than average gas production and had to consider the volume of gas generated in relation to the size of the electrolyzer tank.

Hydrostatic pressure

The gas pressure for the electrolyzer system is provided by a hydrostatic column. This can be simply a vertical tube, the height of which provides pressure to the system as described in this formula:

1 psi = 2.31 ft. water column,

or, to put it another way:

.036127 psi per inch column of water.

Thus, a tube 5′ high would provide a pressure of 2.17 psi, a tube 6′ high would provide a pressure of 2.60 psi, a tube 7′ high would provide a pressure of 3.03 psi, a tube 8′ high would provide a pressure of 3.47 psi, and so on.

The hydrostatic column is a simple way to regulate the pressure up to a certain point. The Romans used this principle to apply pressure to the water supplied through pipes from their aqueducts This method is suitable and cost effective for stationary systems that require only a small amount of pressure to operate. It is also somewhat safer than pressurized systems when a caustic electrolyte is used.

An enclosed pressurized system is the next step for experimenters who wish to design a more compact system, such as the system in my book *Practical Hydrogen Systems.* Compactness can be achieved by designing the electrolyzer system to withstand higher operating pressures, as you require. Such a system needs the addition of relief valves to guard against over pressure, and a pressure gauge and float valve to feed the electrolyzer with KOH solution or distilled water. The hoses in the system also have to have sufficient rating for the pressure, and require hose band clamps, and so on.

To keep things simple and inexpensive for this project, we used the hydrostatic column to provide gas pressure to the system. This low pressure approach permits more flexibility in that components can be quickly added or subtracted to refine or change the system design as a whole. The components are also less expensive, which is useful for building prototypes and learning the basic principles of operation. Once you get used to their operating quirks and understand how these systems operate by observation, then you can upgrade them, and graduate to more expensive components – and make pressure design changes with some experience behind you. This more studied and slower approach will give you the skills to go on and design some very compact and sophisticated systems.

Building the P41 Electrolyzer

Tool list

All should be available from local hardware or electronics stores.

Long nose pliers

Epoxy glue

Scissors

Long screwdriver, 6" or more

Metal cutters

Soldering iron, 15 watt, pointed tip

Caulking gun

Screwdriver

Drill bits, 1/4" and 3/8"

Hacksaw, and other types of saws to cut metal and other items

Drill, either handheld or drill press

Materials list

All should be available from local hardware stores , unless otherwise noted.

11 Stainless steel nuts for 10/24 screws.

1 PVC pipe, thin wall 3" ID, 9 1/8" length.

Nickel alloy foil tabs, 0.01 thick. 5 pieces each 1/2" to 3/4" long, and nickel alloy washers 1" diameter. McMaster-Carr, part #8912K24, comes in a sheet 4"x 12".

Polypropylene white felt, 1/16" thick, 7 3/4" long x 5 15/16" width. McMaster-Carr, part #88125K11. Rolls are 12"x 72".

1 PVC pipe coupler 1 15/16" ID, 2 7/16" long.

1 PVC pipe cap 1" ID.

6 Stainless steel washers, 1/2" diameter to accommodate 10/24 screws.

4 Silicone rubber spacers 1/8" thick x 1/4" wide X 6" long. 2 silicone rubber washers 1" diameter. McMaster-Carr, part #7665K22, comes in a 1" wide strip, 36" long.

Materials list (continued)

1 Rubber O-ring 1"ID.

Screws – pan head stainless steel, two 10/24 x 1", and one 10/24 x 1 1/4".

Monel standard grade woven wire cloth, 200x 200 mesh, .0021" wire diameter, 12"x 12" sheet. McMaster-Carr, part #9225T361.

2 PVC pipe caps for 3" thin wall PVC pipe.

Barbed hose connectors, two 3/8" and one 1 1/4".

Clear silicone rubber caulking.

Parts for the P41 electrolyzer, felt separator and pipe coupler not shown.

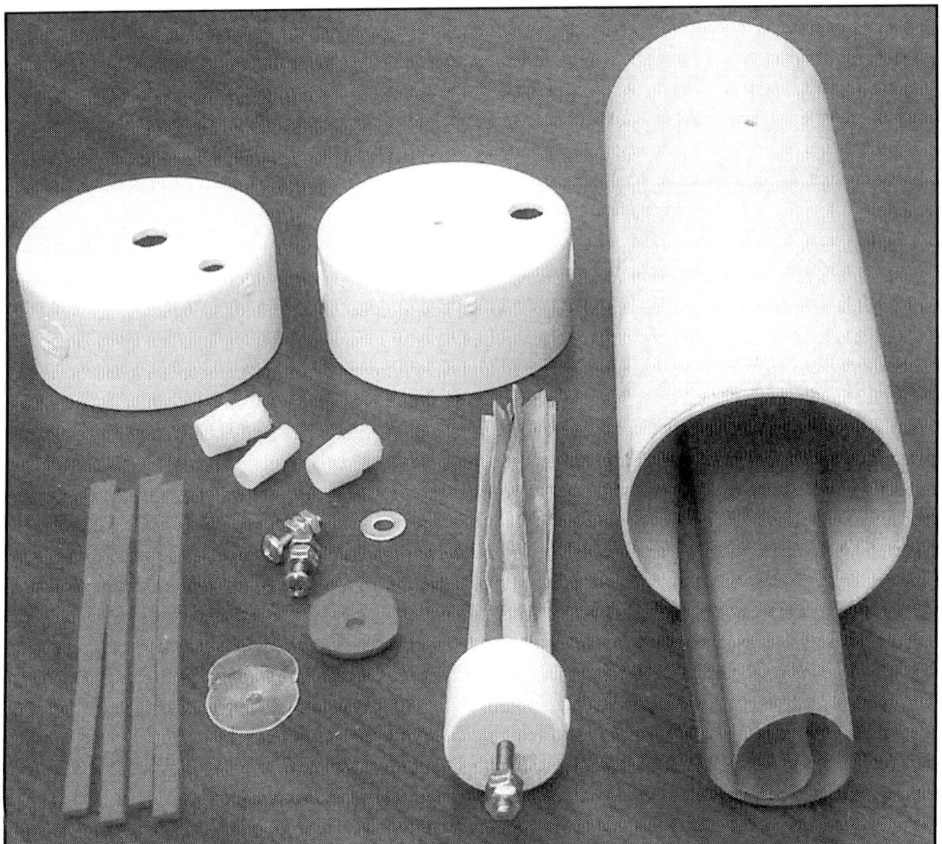

Building the P41

The P41 consists of very few parts. It is inexpensive to build, is suitably matched for renewable energy inputs from photovoltaic panels or wind generators, and it has a high volume of gas output for its size.

Electrolyzer tank

The electrolyzer tank is easily constructed from thin wall PVC pipe and pipe caps. Cut one piece of 3" thin wall PVC pipe to $9^1/_8$" length. This will be the body of the electrolyzer chamber.

Positive electrode assembly ports

Drill a hole in the $9^1/_8$" pipe to accept a 10/24 screw 3" from one end of the pipe. Drill a similar size hole directly opposite that hole on the other side of the pipe. You will then have two holes drilled to accept 10/24 screws on opposite sides of the pipe that are 3" from one of the pipe ends. These holes are the entry ports for the positive electrode assembly.

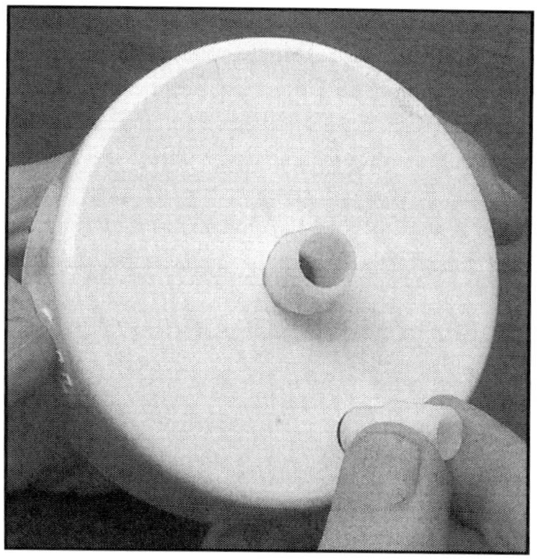

Prepare the gas port cap

Gas exit port cap preparation

Take one of the 3" pipe caps and drill a $^3/_8$" hole in the center of the cap. The hole is the hydrogen gas exit port, and the cap will be the top of the electrolyzer. Center a $1^{15}/_{16}$" pipe coupler inside the cap (see bottom photo at right), and use a fine tip

marking pen to outline the coupler's outer circumference inside the PVC cap. This circle is a gluing guide for attaching the coupler to the cap.

In the same cap, drill another 1/4" hole between the inner wall of the pipe cap and the circle outline you drew in pencil. This hole is the oxygen gas exit port.

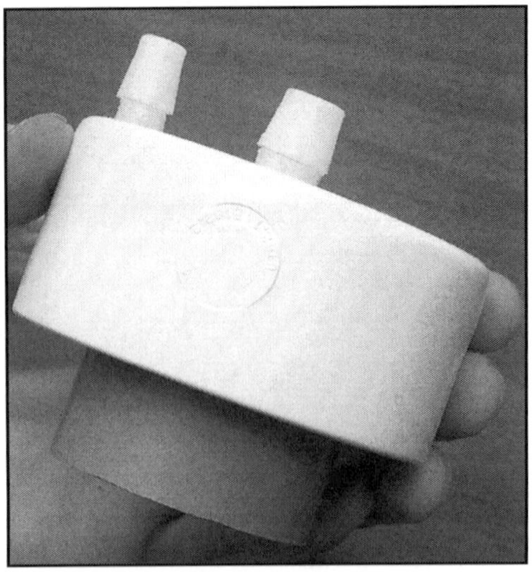

Side view of the gas port cap

Take a 1 15/16" ID (inner diameter) x 2 7/16" long PVC pipe coupler, and coat one edge with epoxy. Coat the circle outline on the inside center of the cap with epoxy also. Align the coupler with the outline of the circle and press it. Put extra epoxy around the outside and inside edges of the 1 15/16" gas takeoff to ensure a very good seal. A tight seal is critical so that the hydrogen does not leak to the oxygen collection side, and visa versa.

Let the cap and coupler dry for 24 hours. Then, apply epoxy around the barb and around the entry port hole, and place the 3/8" cut off barb in the center. The barb should not protrude into the cap. It should be flush with the inside wall. Leave the barb long enough on the top, outside of the cap, to have as much as possible to fit the hose over. If it is too short, the tube will not have a good gripping area when it is placed on the barb. Repeat the same process for the oxygen take off. Let it dry for 24 hours.

Be careful drilling holes for exit barbs in these thin walled pipe caps. If you drill them a tad too large or at a slight angle, the barb will be floppy when seated on the cap. The fit should be as tight as possible.

Separator preparation

Cut a piece of polypropylene felt 7$^{3}/_{4}$" X 5$^{15}/_{16}$". Set up a 15 watt soldering iron with a new tip. I used a pointed tip, but you could experiment with what are called chisel or screwdriver type tips, or try a higher wattage iron and see how that works. A heat gun or wood burning set element might work quite well also.

Use the soldering iron as a heat welder to bind the edges of the felt together to form a tube. The tube will be 7$^{3}/_{4}$" long. The circumference of the tube will be formed from the 5$^{15}/_{16}$" width of the felt. To form the circumference, fold two edges to the center of the width and make sure the edges abut well and evenly (see below). Hold the edges together on one end and apply heat to bond, by stroking the soldering iron across the seam where the two edges abut. This will melt the plastic and bind them together.

Use a soldering iron to bond the seam of the rolled up felt

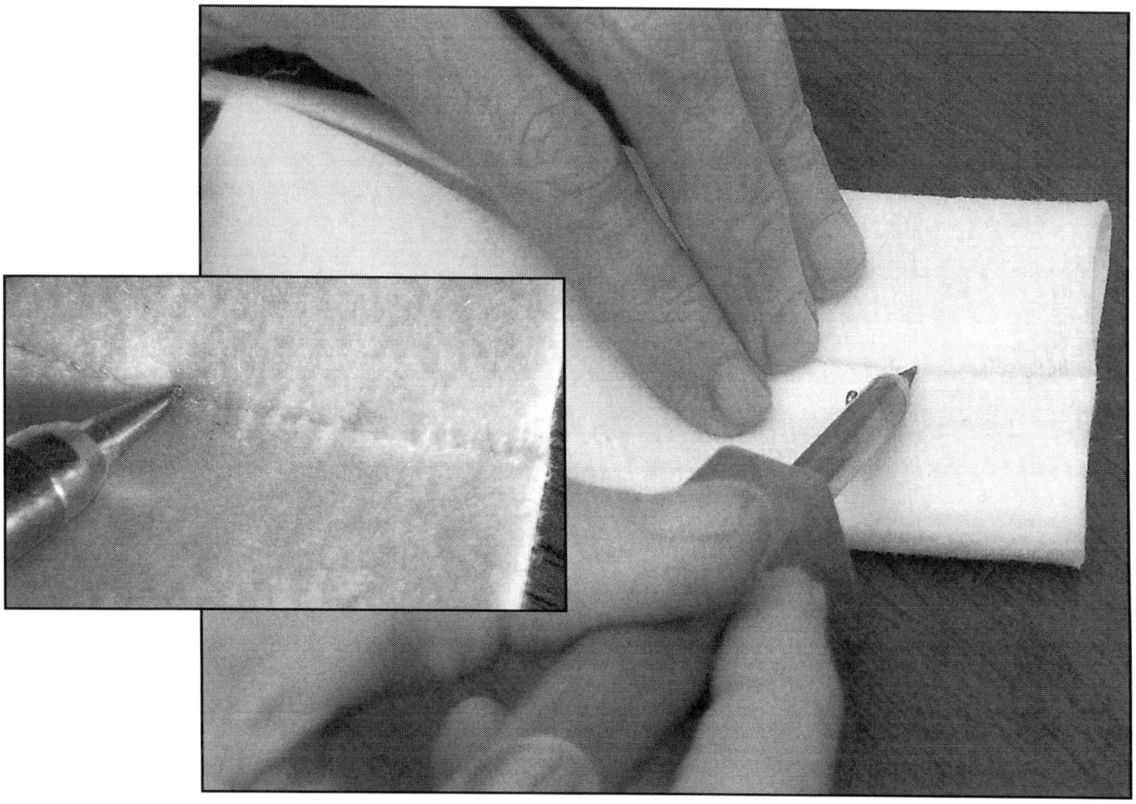

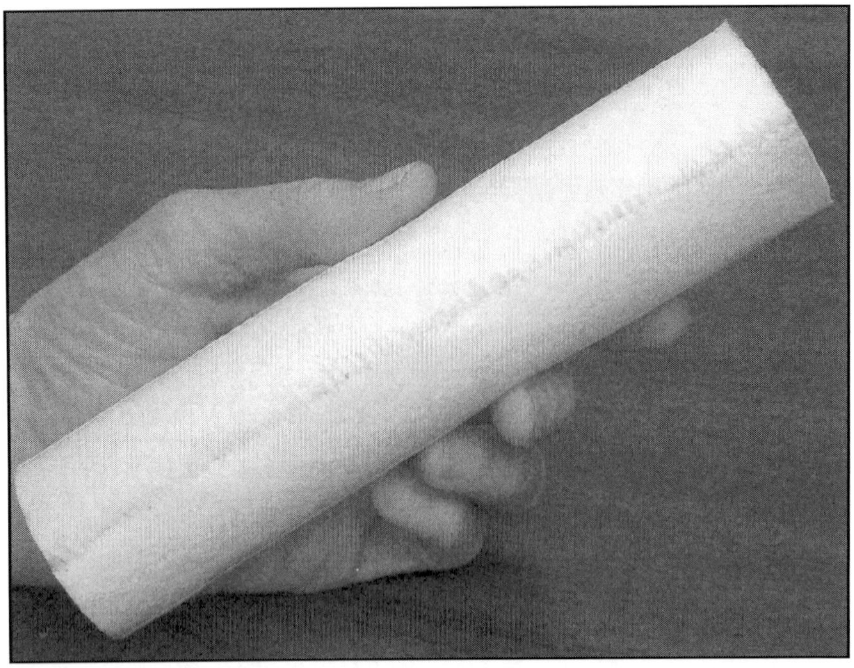

Separator tube after bonding

Move down the piece and stroke across the edges as you go, welding the entire length of the tube in this manner. Stroke the iron quite quickly and lightly over the seam. If you hold the iron a second too long, you will melt too much plastic and ruin the piece. Be sure to make each welding stroke close to the previous one so that there are no gaps in the weld. If you hold the piece of felt up to the light you will be able to see any gaps. If there are gaps, go over the gap area lightly until it is closed. It is important to make sure that no gaps exist so that the hydrogen and oxygen gases do not mix. Practice the technique on some scrap pieces first, and then proceed to the final piece. This felt tube fits inside the 1 15/16" pipe coupler on the hydrogen gas exit port.

Positive electrode assembly preparation

Cut a 6"x 9 1/2" inch piece from the Monel wire cloth according to the layout on page 88. This will be the outer oxygen electrode and will follow closely the inside surface of the PVC pipe.

Glue rubber spacers on the inside of the electrolyzer tube

Cut four silicone rubber rib spacers 1/8" thick, by 1/4" wide, by 6" long. Cement these strips with silicone or epoxy to the inner walls of the electrolyzer chamber more or less equidistant from each other, with one end of the strip lining up with the bottom edge of the pipe. Be sure not to cover over the positive electrode port holes. These strips keep the wire mesh away from the wall of the pipe so that gas can be created on the outward side of the positive electrode.

Washers

The next step is to make washers for the positive electrode assembly. These are two nickel alloy washers and two silicone rubber washers. Use a 1" diameter washer as a model to trace the circumference and center hole onto the rubber and the nickel alloy sheet. Use a punch to cut out the center hole in the rubber sheet, and a punch or drill to cut out the center hole on the nickel alloy. The circumference of the rubber sheet can be cut with an exacto knife and the circumference of the nickel washer can be cut with tin snips.

Secure the rolled screen at the correct diameter with crimped nickel tabs

After the rib spacers' glue has set, roll the 6"x 9 1/2" monel mesh screen into a cylinder (6" long) that can be slipped into the electrolyzer body. When the piece of mesh is fully inside the tube on the ribs, adjust it to even it out. There should be an overlap. This is the tabbing junction. Adjust the monel screen so that it is even, round and sitting on the ribs well, but not touching the walls of the pipe. This is the proper circumference to allow gas flow between the mesh and the walls of the pipe.

Cut crimp tabs from the nickel alloy. These crimp tabs secure the overlapping edges of the Monel mesh on both ends of the cylinder (see illustrations above and below). Cut the crimp tabs about 3/4" long and about 1/8" wide with tin snips. They can be wider and longer if desired. Make sure that the mesh cylinder retains its form; then, slip the cylinder a little bit out of the pipe and use pliers to crimp the tab on

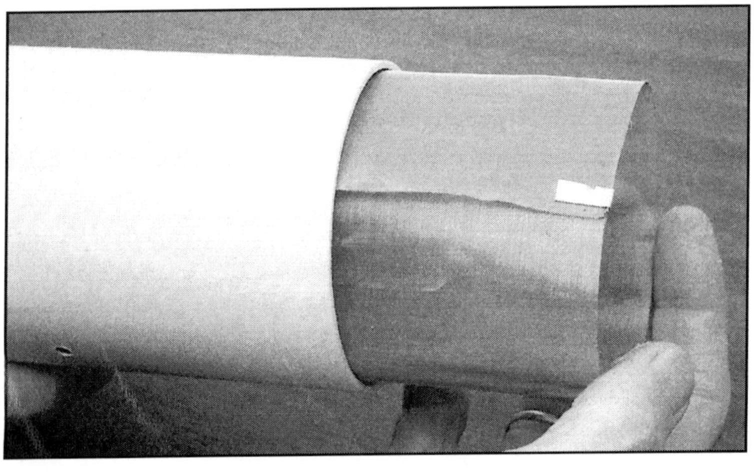

Slip the cylinder back into the pipe

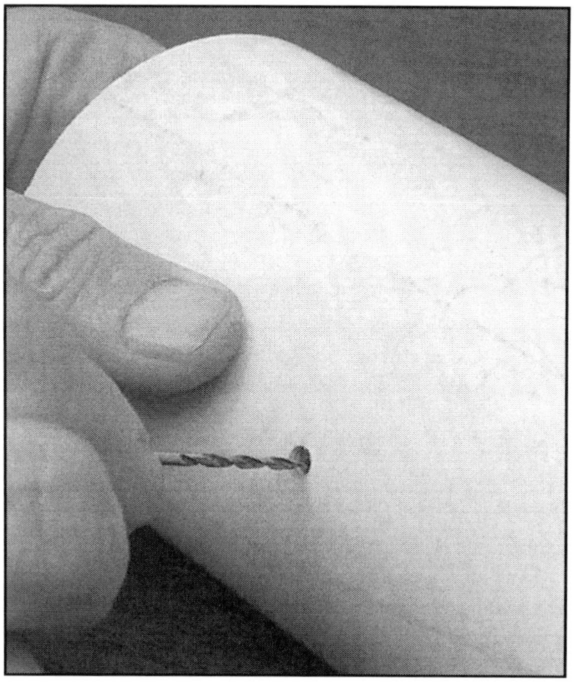

Insert a drill bit through the hole in the pipe wall to make a hole in the positive electrode screen

the overlapping mesh to secure the two edges together. Pull the mesh totally out of the tube and secure the other end of the cylinder with a crimp tab.

Slip the cylinder back into the pipe and position the overlapping seam over one of the electrode port holes. Inside the tube, place your fingers over that hole, and press the mesh against the pipe.

Insert a pointed object such as a small drill bit through the hole from the outside and push it through the mesh. The drill bit will move the mesh strands aside to create the hole rather than breaking them. This hole needs to accommodate a 10/24 screw that will be inserted through it, so be sure it is the correct size. Remove the drill bit and replace it momentarily with the 10/24 screw to hold the mesh in position while you make the hole on the opposite side of the cylinder. Create this second hole the same way as the first. Take care doing this so that the cylinder retains its shape and the holes are correctly aligned.

Slip out the screw and the mesh cylinder.

Take a nickel alloy washer, place it on a 10/24 screw and insert the screw through one of the holes in the mesh cylinder, with the head of the screw and the nickel washer on the inside of the cylinder. Then slip a silicone rubber washer on the outside of the cylinder over the screw.

Do the same with the other port hole. Back off the screws a bit to fit the cylinder back into the pipe, then push the screws through the portholes so that they extend to the outside of the PVC pipe.

The next step is to screw on a nut. Apply a good quantity of silicone to seal the hole and the nut to the hole and pipe before you tighten the nut to the pipe wall. Tighten the nut and apply more silicone to make sure the nut is sealed well against the hole. Repeat for the other hole.

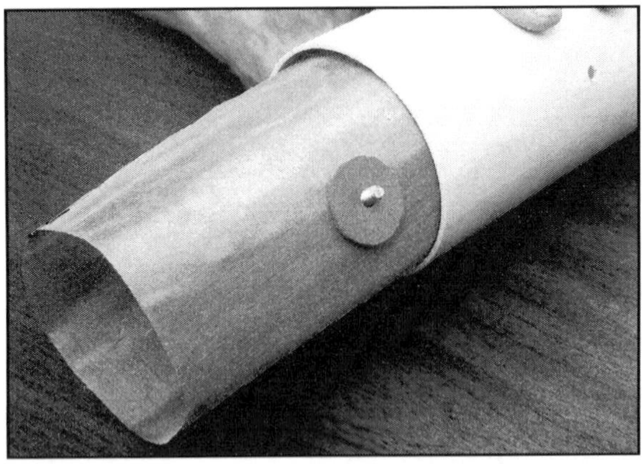

Insert the electrode connector screws from the inside, though a nickel washer, then the Monel screen, then a rubber washer before they go through the pipe wall.

During this process, try to be neat and not get silicone on the screw surfaces where the electrical connections will

be made, as the silicone will insulate the electrode from the power source and the current will not flow. If you mistakenly get some on the screw body where the contacts will be, clean it off thoroughly to be sure the surfaces are electrically conductive and not impeded by silicone.

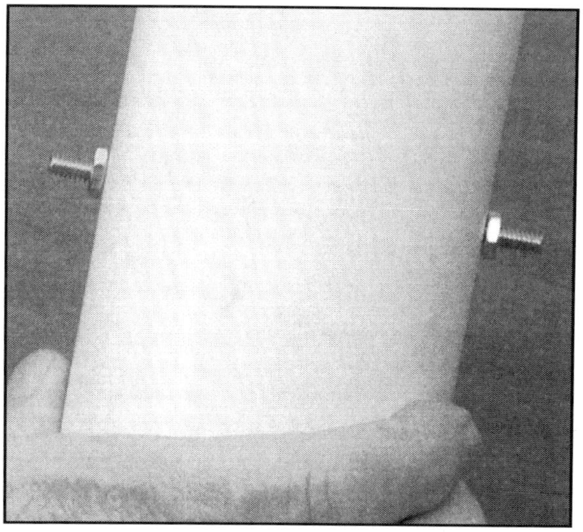

The installed positive electrode connectors

When the silicone has dried thoroughly, slip another nut on, and tighten it against the first nut, then slip the two washers on and the other nut. The connecter will be attached to the electrode assembly between the two washers.

Finally, test for continuity with a multimeter. If you are not getting continuity, check to see if there is silicone between the pan head of the screw and the washer and screen. As mentioned above, silicone will insulate the connection and the current will not flow. If you apply glue as neatly as possible in the first place, there shouldn't be any problems.

Negative electrode assembly preparation

Cut the monel wire mesh for the negative electrode to dimensions as shown in the diagram on the next page. This will give you a 6"x 9 3/8" piece with a 3/4" wide tab at the bottom. This piece of mesh will be folded into pleats.

To make the pleats, mark the fold points on each side of the screen (see illustrations, page 88-89). The fold points should be 3/8" apart. Begin folding, using a ruler along each fold line as shown in the illustrations, pages 89-90.

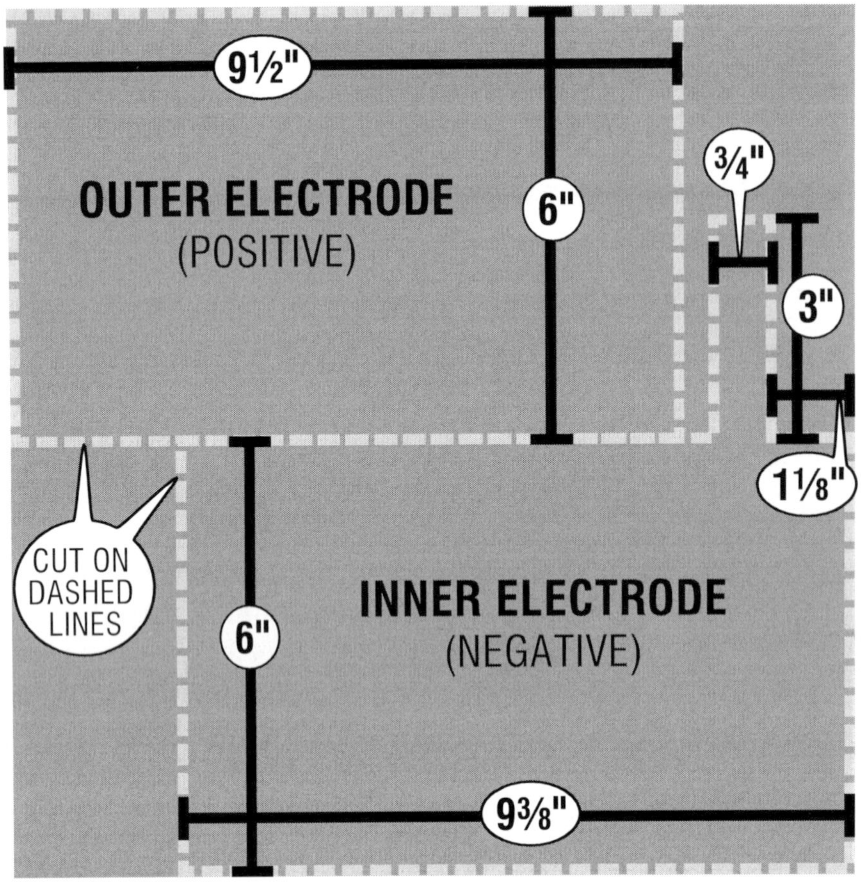

Above, layout for cutting the positive and negative electrodes from a 12"x 12" piece of monel screen

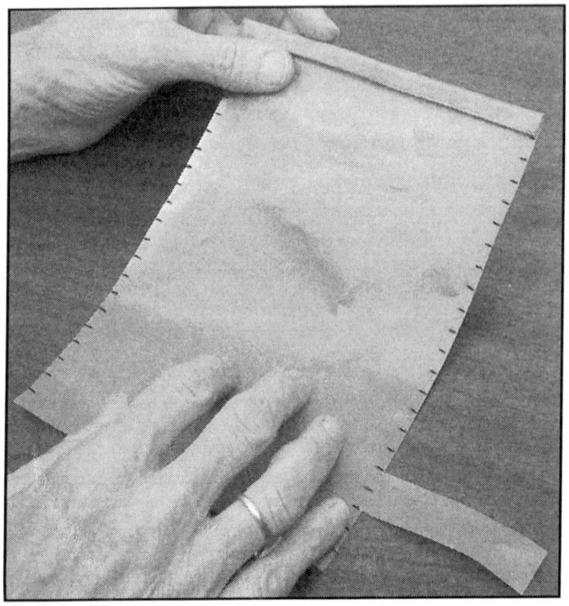

Mark the fold lines for the pleats on the screen, as shown above right, next page.

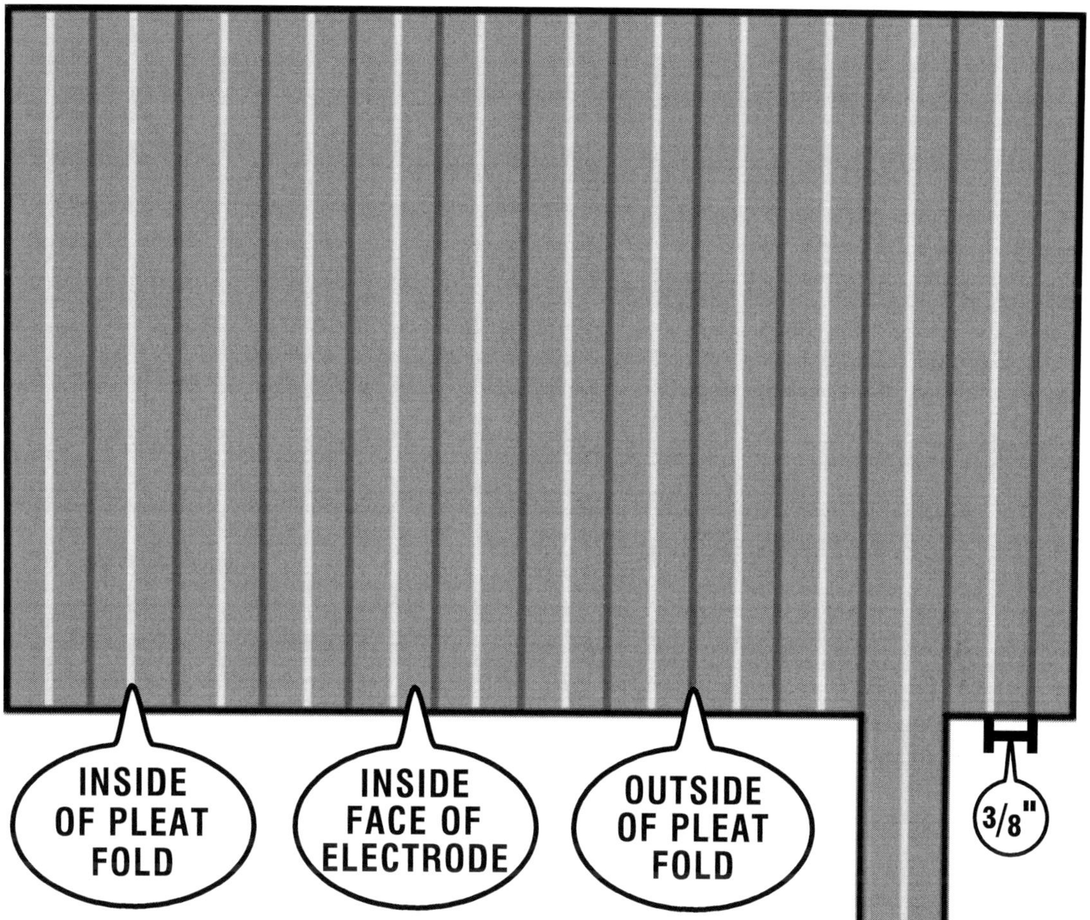

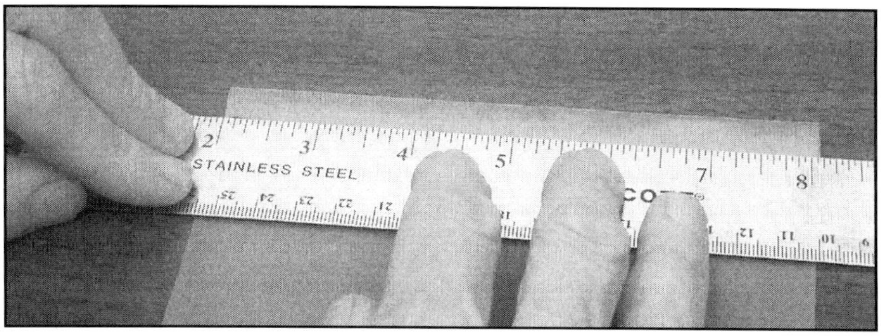

Use a metal ruler to make sharp creases for each fold of the pleats

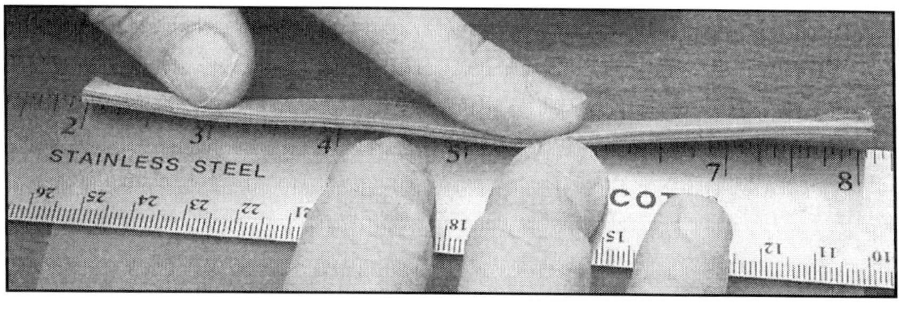

Right and below, folding sequence for pleating the negative electrode.

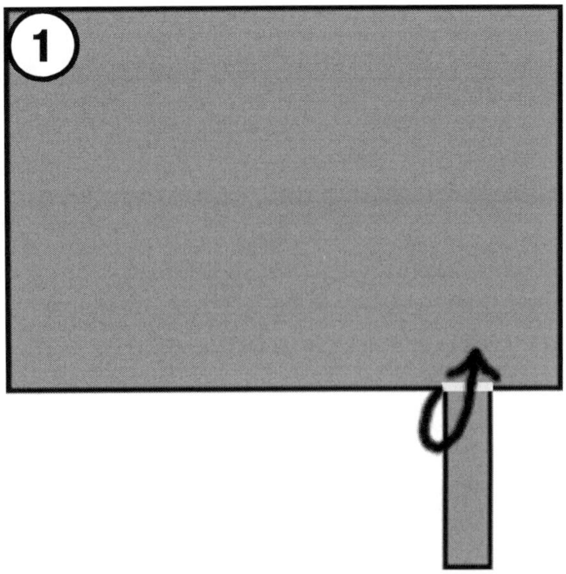

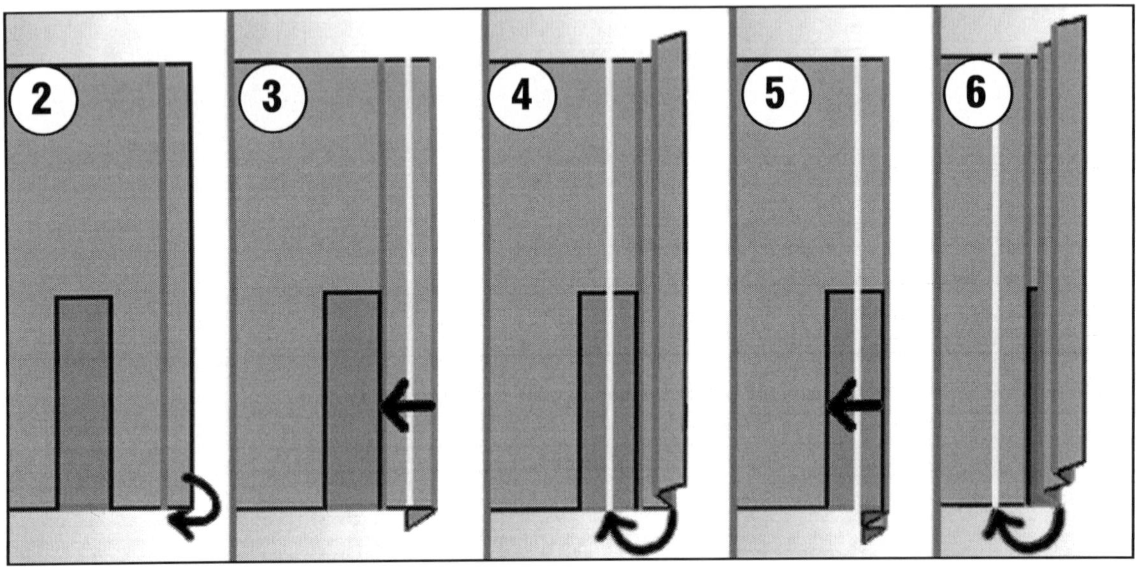

When all the folds/pleats have been made, fold the tab back a long the edge of the main piece towards the pleats and crease the fold, as shown at right, top (step 1), then finish the tab by folding as shown in steps 2-4, next page. Roll the ends of the pleated mesh together with the tab coming out from the inside of the roll, overlap the edges as shown at left, then crimp the overlapped end pleats together with nickel alloy tabs. With a drill bit or other small pointed object, make a hole in the center of the tab for a 10/24 screw (step 4, next page)

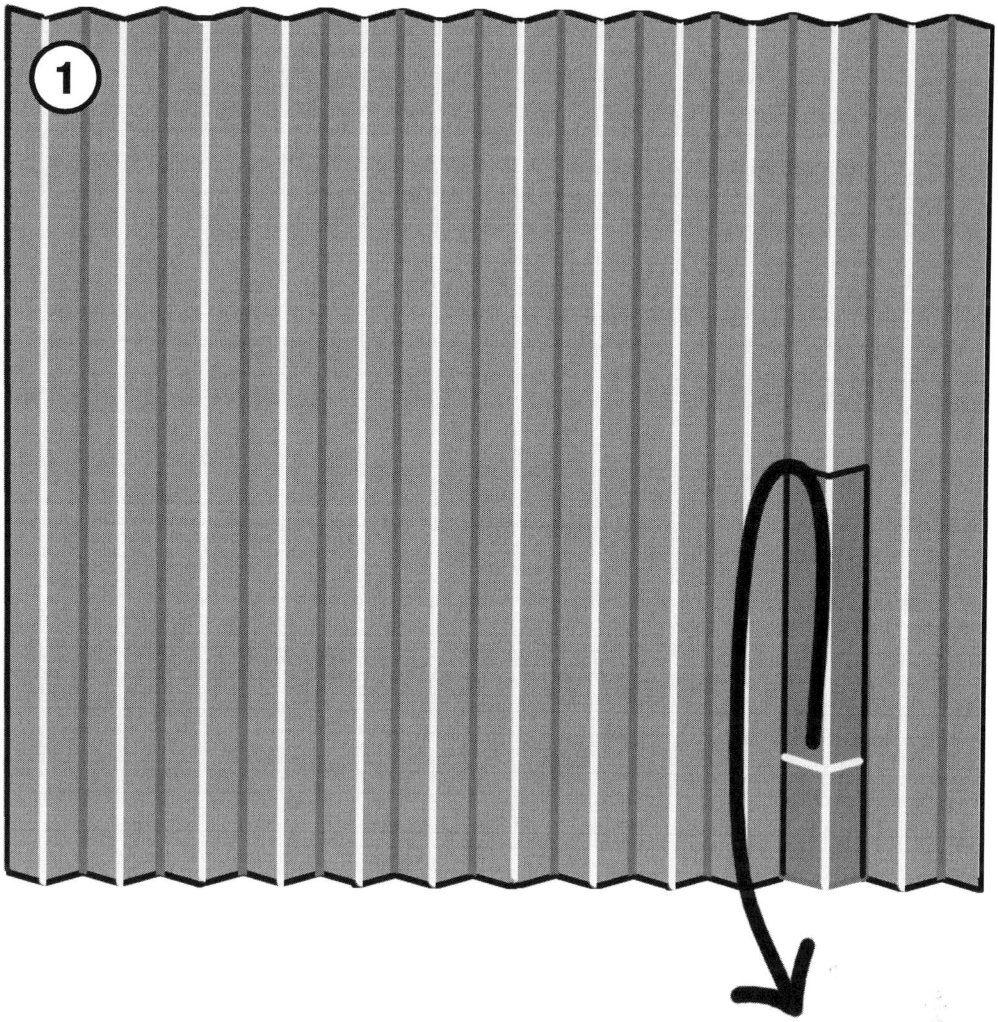

Folding sequence for negative electrode tab connector

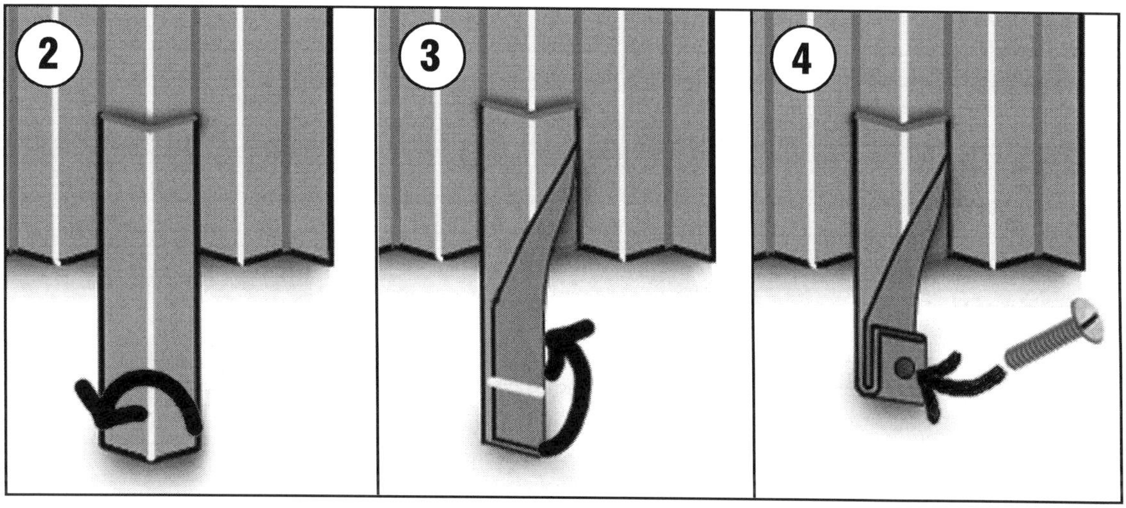

Roll the pleated electrode and overlap the edges as shown, then crimp with nickel alloy tabs to hold in place.

The pleated and rolled negative electrode, ready to install in the bottom cap.

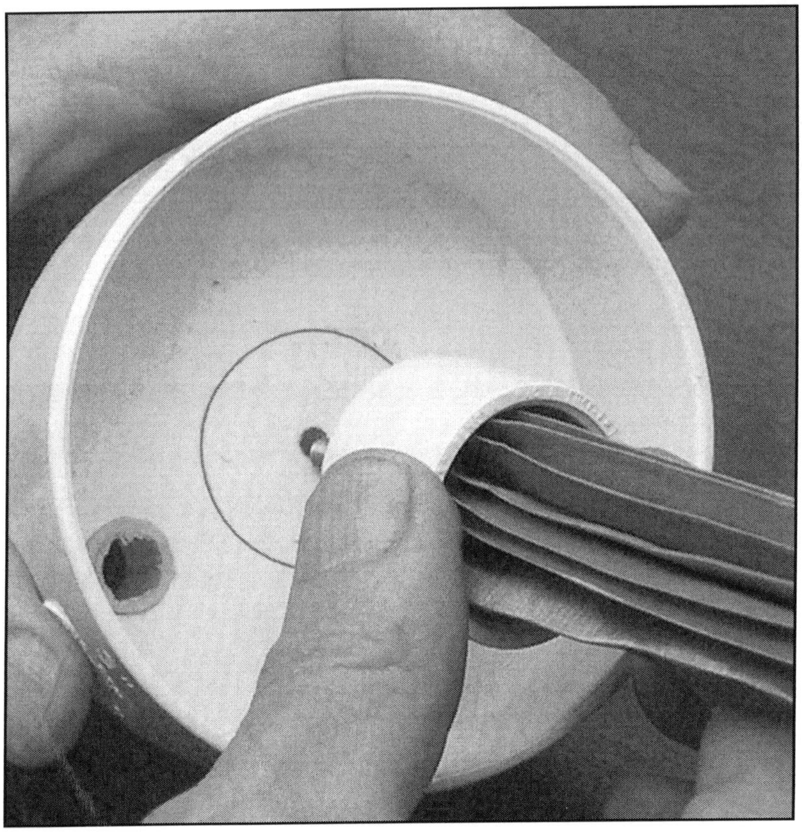

Inside the bottom cap, showing guide outline, center screw hole, and electrolyte entry port

Preparing the bottom cap

Drill a hole in the center of the pipe cap for the bottom cap to accept a 10/24 screw. Then, center the 1" PVC cap inside the cap, and draw an outline with pencil. The 10/24 screw hole should be dead center within this outline. Drill a 3/8" hole between the inner wall of the 3" PVC cap and the circle outlined for the 1" PVC cap (see illustration at bottom left). This 3/8" hole is the electrolyte entry port. Cut and epoxy the 3/8" barb connector into the hole.

Drill a hole in the center of the 1" inner diameter (ID) PVC cap to accommodate a 10/24 screw. Make a 1" nickel alloy washer. Apply a small amount of silicone around the area outlined for the 1" ID pipe cap on the inside of the 3" pipe cap. Keep the silicone away from the center hole which will accommodate the 10/24 electrode connector screw.

Apply silicone to the outside top of the 1" ID pipe cap. Keep the silicone away from the electrode hole so that when the screw connector goes through this hole, it does not get silicone on its threads, which would compromise the electrical connection. Now, push the 10/24 screw electrode connector through the wire mesh (see illustration)

Assembling the connection for the negative electrode

Place the 1" ID cap in position in the center of 3" electrolyte entry cap and seat the silicone covered surfaces. Drop the 1" nickel alloy washer into the 1" ID cap. It should fit in snugly.

If the entry holes for the 1" pipe cap and 3" pipe cap are tight for the screw, start the screw in and then back it out. This will cut threads into the pipe. If the holes are not that tight the screw will push through without threading. Ideally, the screw should go in tightly.

Take a 6" long screwdriver and insert it down through the center of the wire mesh electrode. Push or screw the 10/24 electrode connector through the holes with the tab mesh connected, as shown. This will seat the mesh tab onto the nickel alloy washer, and secure the mesh electrode to the cap. Apply silicone around the screw and screw hole where it exits the cap. Use enough silicone to form a good seal for the nut. Put on a nut and tighten it. Apply more silicone around the nut where it touches the PVC, again being careful not to get silicone on the part of the screw above the nut, as it will interfere with a good conductive connection. Inside the cap, apply more silicone where the

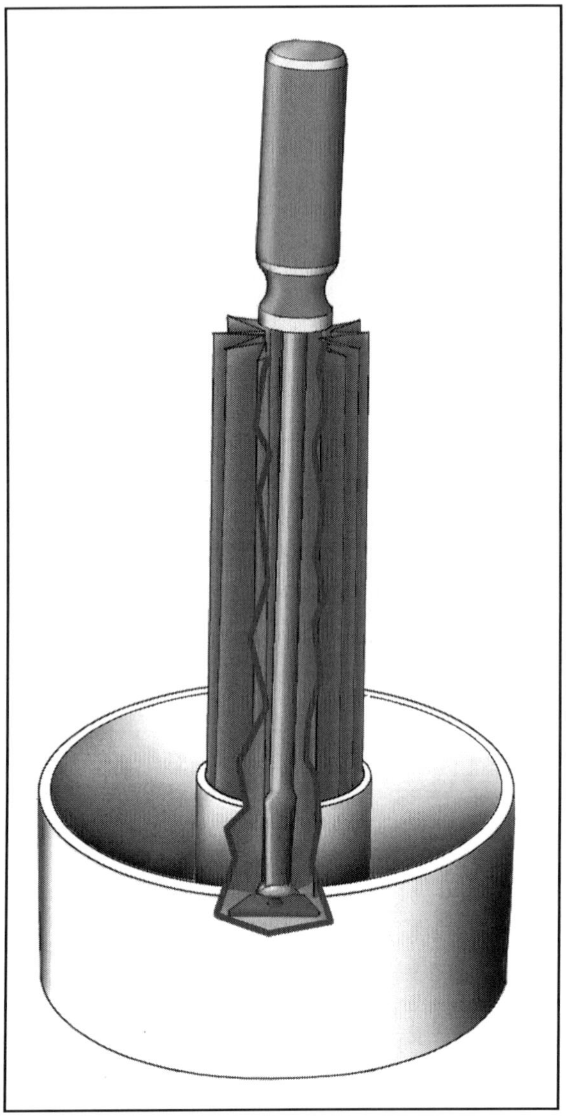

Use a long shafted screw driver for the electrode connection screw.

1" cap comes together with the 3" cap. You can smooth this seal with your fingers.

The next step is to put some silicone around the rim of the 1" cap that seats the negative mesh electrode. Take a 1" rubber o-ring, slip it over the mesh and slide it down to the 1" cap, seating it on the cap. The rubber o-ring will steady and keep the mesh electrode straight. It is important for the mesh electrode to stand straight up and not be seated at an angle.

Let the silicone dry. Test the mesh electrode for looseness in the 1" cap. If it stands straight up it is ok. If not, apply a drop of epoxy where a pleat or two touches the 1" PVC pipe and let it dry. While drying, make sure the mesh screen stays straight and vertical.

Installed positive and negative electrodes.

Installing electrolyte entry head/negative electrode assembly

Coat the outside of the pipe and inside of the top pipe cap with silicone, as done with the entry head/bottom cap. Be sure that the negative electrode is secure and standing straight upward. Take the felt separator tube and slip it into the

1 15/16" hydrogen exit port attached to the exit head. Then, slip the other end of mounted felt separator tube down over the negative electrode, and seat the cap, finally enclosing the electrolyzer chamber. Coat the seam between the cap and the electrolyzer chamber tube with silicone where they meet and smooth out with your finger.

Installing gas port exit cylinder head and gas separator

Coat the outside top edge of the pipe with silicone around the circumference for the length that the pipe cap will travel when seated. This will be about 1 1/2". Coat it liberally. To form a gas seal, the area must be thoroughly covered. Then, coat the rim of the pipe with silicone, and then coat the pipe cap walls with silicone around the inner circumference. Make sure the surfaces are well covered.

Do not use so much silicone that it will be messy and squeeze out all over the place, but at the same time be sure that the coverage is thorough to make a good gas seal.

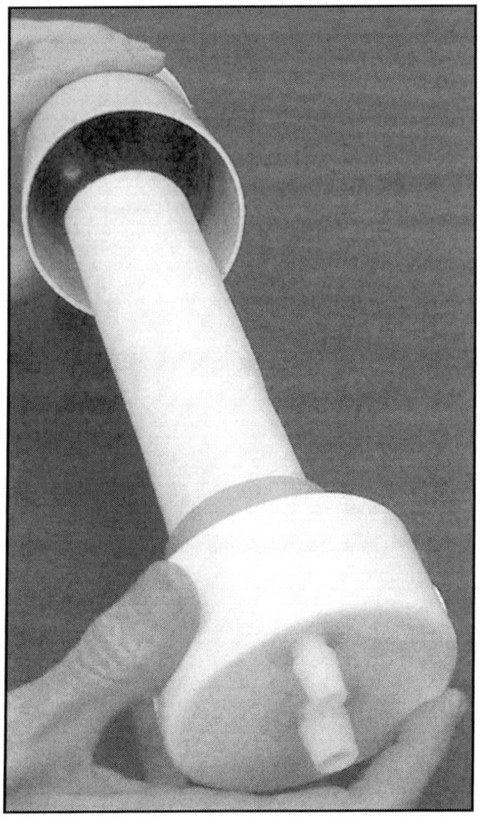

Seat the pipe cap, sliding the felt tube down over the negative electrode. Make sure the cap is pressed all the way down. Use a flashlight to see if any silicone is impeding the oxygen port. If it is, remove it with a long stick. I use wooden BBQ sticks from the grocery store. They are handy for gluing in long tight spaces.

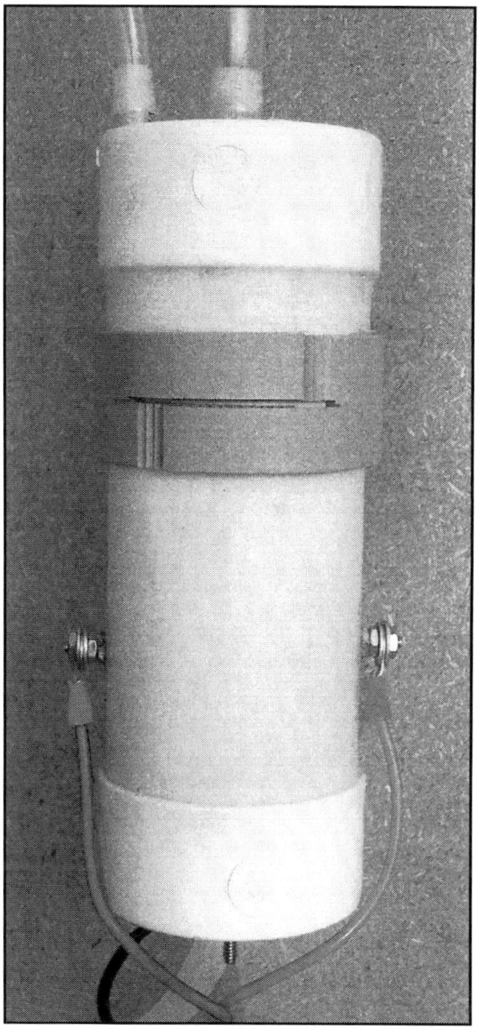

The completed P41 electrolyzer

Congratulations, you have just constructed a very good quality, high performance electrolyzer!

Hydrostatic testing

To test the integrity of the silicone bonding of the electrolyzer's components, conduct a few leak tests. Use a Rubbermaid® container or a bucket filled with water. Connect pieces of hose to each of the ports (hydrogen outlet, oxygen outlet and electrolyte inlet). The hose pieces should be long enough so that you can immerse the electrolyzer completely in the water and have the open tube ends above the water with no liquid flowing into the electrolyzer through any of the three ports. Block off the ends of two of the tubes, blow through the third open ended tube and note if any bubbles come from the electrolyzer. If they do, note where the leak is. Take the electrolyzer out of the water and dry it, then seal any leaky spots with more silicone. When the silicone dries, leak test it again, note any further leaks and apply silicone as before until all leaks are corrected.

Power supply connections

The power cord coming from the power supply should be a 12 gauge, or preferably a 10 gauge zip or other wire if you are using

ESPMs. Other wire can be used, as long as it is rated for this use.

The connectors crimped on the end of the wires that connect to the electrolyzer should be ring type connectors so that they will not slide off if the nuts loosen. To connect the ring connector to the electrolyzer, slip one washer against the nut on the electrolyzer tube, slip on the ring connector with wire attached, put another washer on and then tighten a nut over that.

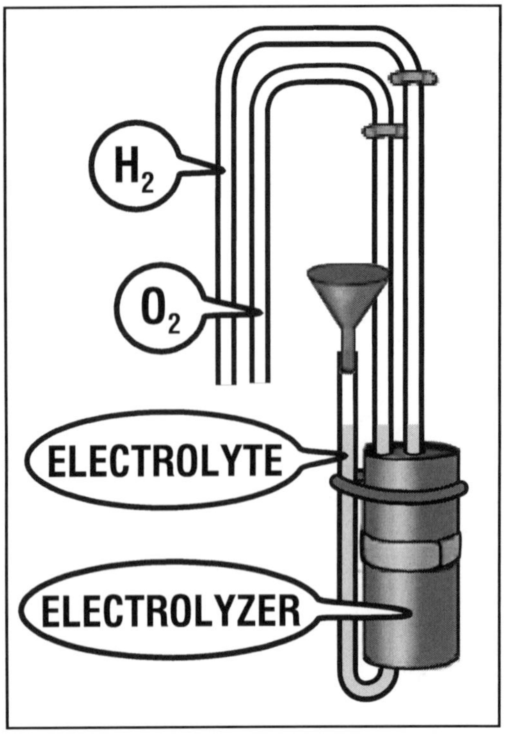

Test set-up

This particular electrolyzer has two positive pole connectors on either side of the body of the electrolyzer, and one negative connection on the bottom. When the wires are connected securely, coat the connectors with silicone to insulate them. You can also use a rubber dip compound or liquid electrical tape for this.

Testing the electrolyzer with KOH

When the electrolyzer has satisfactorily passed the leak tests, connect tubing to the KOH port at the bottom of the electrolyzer. This tubing should be long enough to come up above the top of the electrolyzer and should be secured to the electrolyzer with a rubber band or cord to hold it in place (see illustration). Push a funnel inside the opening of the tube and use this as a fill funnel for your KOH. Attach tubes to the two gas exit ports and support them above the electrolyzer. Fill the electrolyzer with KOH solution through the funnel. Wear protective clothing, gloves, and you absolutely must wear safety glasses or a safety face shield. Fill the electrolyzer just a little above the top so that you can see the KOH solution rising in the tube just above the top of the electrolyzer in all three tubes.

The tubes from the gas ports should be long enough to put them in a glass of water. Connect the electrolyzer to the power source and wait for the bubbling action to start in the glass. If everything appears to be working, empty the electrolyzer carefully and the testing is complete.

You can run distilled water through the electrolyzer after testing and emptying out the KOH, to flush out any residual KOH solution.

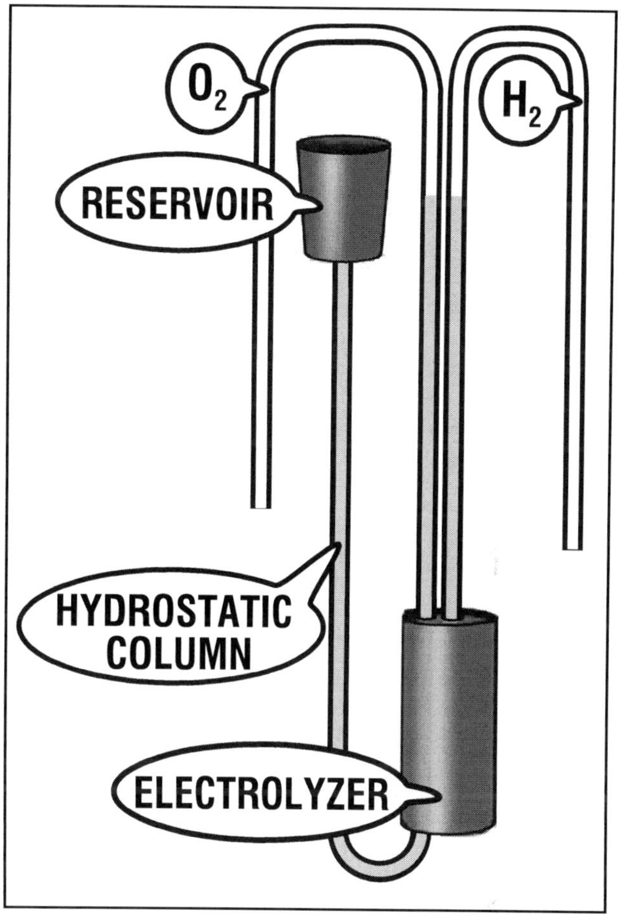

Hydrostatic column set-up for operation

Setting up the electrolyzer

Level the electrolyzer for operation. Attach the hydrostatic column/electrolyte feed tube to the feed port at the bottom of the electrolyzer. The length of the feed tube/column will depend on how much pressure you want. Any length from 3.5 feet to 8 feet will suffice. You can add a set of gas check loops to connect the hydrogen, oxygen and hydrostatic column tubes together (see illustration next page). This

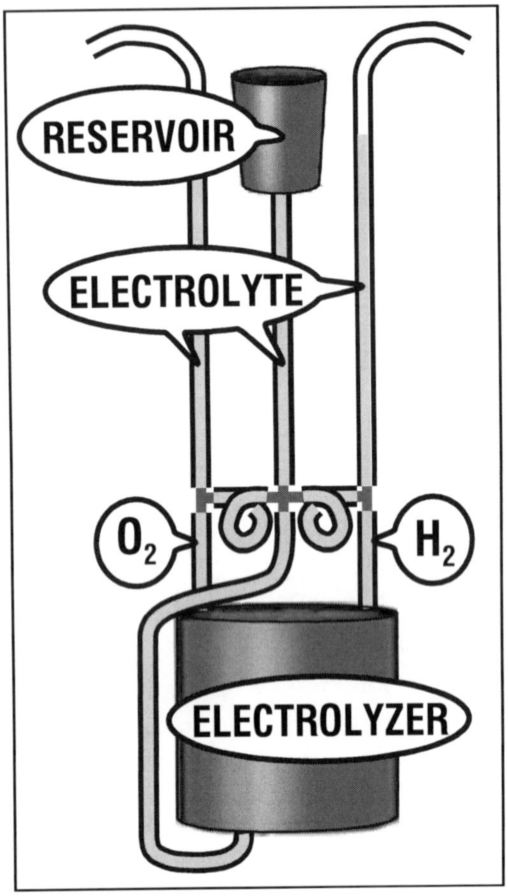

Hydrostatic column set up with gas check loops

is not necessary, but if the electrolyte runs low in the tubing, this will give the gases an outlet through the hydrostatic column.

Set up the hydrostatic column along a wall or other support structure, and attach the tubes with pipe hangers. Attach the top of the column tube to a reservoir.

Electrolyte reservoir

The reservoir will be filled on a regular basis to maintain hydrostatic pressure and provide electrolyte and/or distilled water to the system. The reservoir should have a cover or lid to keep out debris and to slow evaporation. Make a small hole in the cover so that a vacuum is not created in the reservoir.

Electrolyte reservoir

The reservoir can be anything from a 32 oz. recycled yogurt container to a large custom built tank. Rubbermaid® containers, and small tanks from Aquatic Eco Systems made from polypropylene or polyethylene are good candidates. You can also purchase one-piece containers with sealed partitions so that you can have several reservoirs in one container. Round containers can be hung on a wall/

support structure with metal flower pot holders or large size pipe hangers. Make sure to use materials that are compatible with KOH. Do not use metal containers unless they are stainless steel.

Whatever type of container you use, drill a hole in the bottom, and insert and epoxy a barb or threaded/barb connector through the hole. This connector attaches to the column/feed tube. You may also want to glue a small piece of mesh over the hole in the reservoir to act as a filter. Let the glued connector dry.

Put a hose clamp over the tube before connecting and then slide the hose over the barb and tighten the hose connector. It is important that this hose connection be very secure so that the tube with KOH will not slip off, and rain caustic KOH down upon you.

Set up the exit port tubes

The exit port tubes need to rise at least 1' above the liquid level in the reservoir. The reason for this is that upon startup and initial operation of the electrolyzer, the electrolyte can build up bubbles and foam rising in the gas outlet tubes. If the foam rises over the high point of the tubing and drips down towards the bubbler it can be released by opening the valve on the first bubbler to let the KOH solution flush out. Be very careful if you do this, as pressure is built up in the system and the caustic will spray out. Wear protective clothing and face shield. Be sure the container you are emptying into surrounds the nozzle of the valve. You can also attach a length of hose to the outlet of the valve and then release the KOH to flow down a small length of tubing aimed away from you and into the KOH container. The length of the exit port tube should be a little over twice as long as the hydrostatic tube, so that it can arc back down and connect with the bubblers.

Connect to the power source

Make all your electrical connections to the electrolyzer from the panels. Make sure they are tight so they will not slide. Cover the terminals of the electrolyzer with liquid electrical tape and/or dip rubber compound or silicone caulk.

Fill the electrolyzer

Finally, fill the electrolyzer by pouring KOH into the reservoir. Observe all safety precautions, as mentioned above, for handling KOH. Let the electrolyte fill up the column until half of the reservoir is filled. Put the reservoir cover on. When this is set up, connect the exit port tubing to the bubbler in the gas processing system.

As you like it

The P41 can be made any size you like, and with many variations. Construction materials, method of construction and dimensions, can all be changed to suit your preferences, needs and experimental design ideas. The electrodes can be made larger or smaller, or configured into different shapes. A large bank of these electrolyzers can be built at a very low cost, which was one of our main goals for this design. You may want to build the tanks from the regular schedule 40 PVC pipe which is more sturdy and resistant to temperature change than the thin wall PVC. Of course, many other materials can be used.

Although we used two contacts for the positive terminal contacts for the electrolyzer, four could be used instead, to equalize the potential more evenly around the mesh electrode. Four equipotential contacts may or may not improve performance. We did not try this idea for this project, but it may be worth investigating. Of course, adding more contacts also adds more resistance, and if you try this you might want to gold plate the screws and nuts to reduce resistance.

A bank of electrolyzers can be set up as you would set up a bank of batteries, or solar panels, in a series, or parallel, or series/parallel connection; or, each electrolyzer can have its own dedicated power source. This last option is actually the most energy efficient configuration.

The configuration you choose depends on a number of factors. If you already have BSPMs on hand and wish to use them in a solar hydrogen fuel cell system, it's best to set up a bank of electrolyzers connected in series.

Quick comparisons

We did run a quick comparison test with another electrolyzer. This particular electrolyzer weighed in at about 30 pounds, used sintered nickel plates, and was about three times as large as our P41. The P41 weighs a little more than a pound and was less than half the size of the commercial electrolyzer.

We connected the electrolyzers to matching solar panels and watched the results. The P41 began gassing immediately, whereas the commercial electrolyzer took about a half hour to release its contents into the feed tubes. This fact pointed out design flaws in the commercial electrolyzer.

Basically, in any electrolyzer you want the gas to get out of the reactor tank and away from the electrodes as fast as possible so that the gas does not interfere with the process. It was obvious at first glance that the commercial electrolyzer had gas pockets that had to be filled before the gas would be released. Even after the other electrolyzer was given time to come up to speed, it was visually evident that there was no contest.

The P41 produced twice the gas that the commercial electrolyzer produced. We also noticed that the P41 performed extremely well under intermittent cloud cover as we had expected. Very sharp peaks and troughs in gas production were evident in the commercial electrolyzer, whereas with the P41, gas production was smoother and more consistent. Although we cannot say scientifically that this was a direct result of the thermal flywheel design, the comparison with the conventional electrolyzer demonstrated the anticipated results.

Sharp peaks and troughs were evident in the other electrolyzer with minor atmospheric hindrances that vary from second to second and or minute to minute, such as moisture clouds or dust clouds. These are not perceptible to the naked eye, but never the less affect the power output of the PV panels and thus the gas production from the electrolyzer. Under these conditions, the P41 exhibited a rolling effect with a more consistent gas output, and with full cloud cover it was producing much more gas than the commercial electrolyzer.

BSPMs and electrolyzers

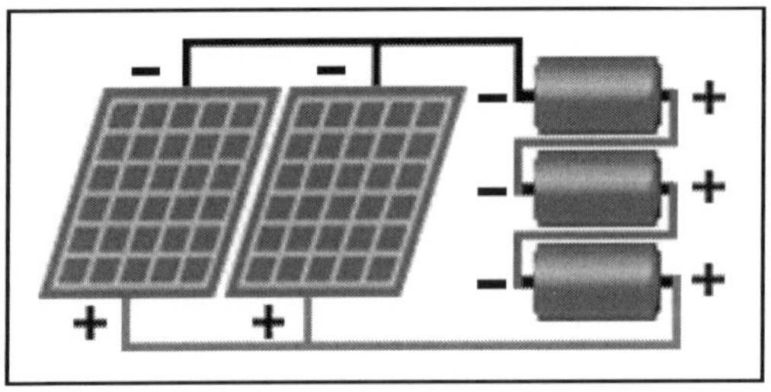

Two 12V 10 amp BSPMs connected in parallel, powering three 4V 20 amp electrolyzers connected in series.

BSPMs (Battery Specific Photovoltaic Modules) give more voltage than current – the average solar panel delivers between 2 amps to 10 amps short circuit current per 16 to 18 open circuit voltage. BSPM panels vary in output but essentially are designed to provide voltage that is above 12 volts in order to charge 12 volt batteries. The voltage of a charging source must be higher than the battery charged in order for charging to occur.

Although such panels are specifically designed for charging batteries, they can be used in systems where a bank of electrolyzers is connected in series. In other words, if you have solar panels that deliver a nominal real working voltage of around 15 volts at 10 amps, you can power about three electrolyzers with two panels, if you want to input 4 volts into each electrolyzer cell at 20 amps, as in the illustration, previous page.

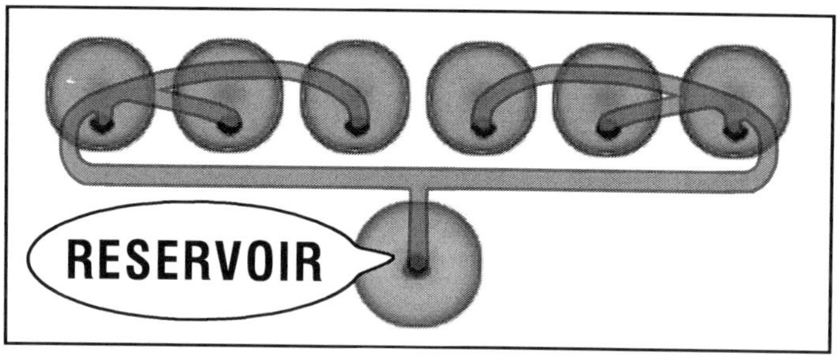

Electrolyte connections for a bank of six series connected electrolyzers with a common electrolyte reservoir.

Series connected electrolyzers

In a series electrolyzer connection, to determine what is needed for solar panel output voltage to power your electrolyzers, add up the voltage that each electrolyzer requires. If you wish to input more current than one panel can supply, connect the two solar panels in parallel, which, if using the same rated panels as just mentioned, will deliver 15 volts at 20 amps.

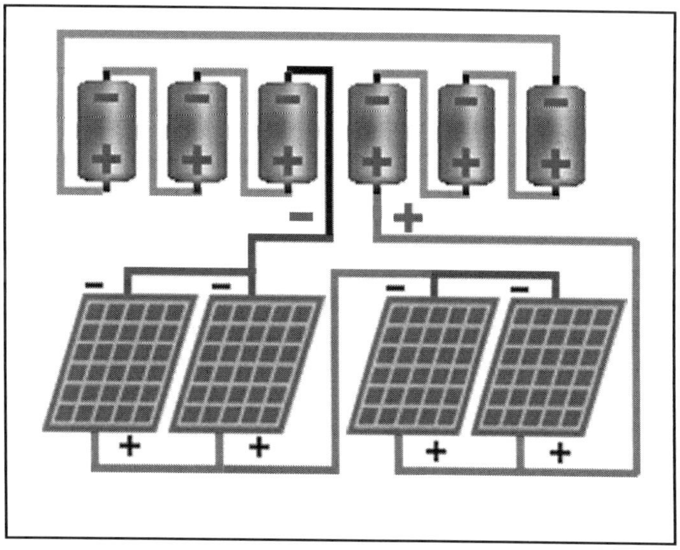

A bank of six series connected electrolyzers with a common electrolyte reservoir, powered by two pairs of parallel connected BSPMs that are connected, pair to pair, in series.

The advantages of using a series connected electrolyzer bank is that you can use a common

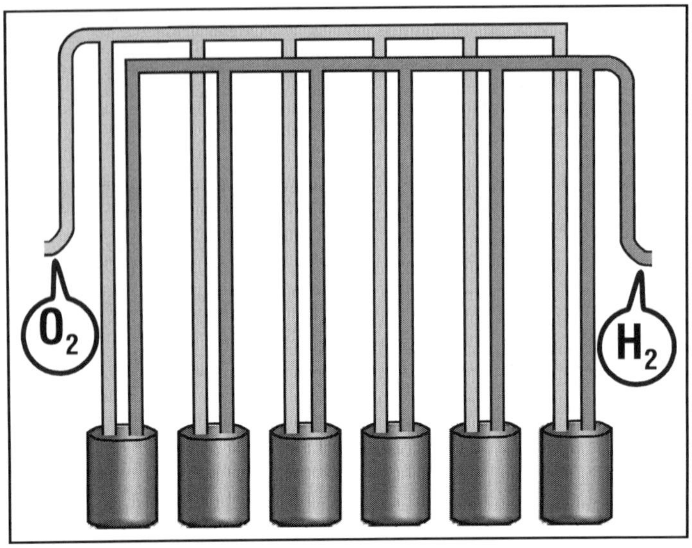

Configuration for gas exit tubes for a bank of six electrolyzers.

water or electrolyte reservoir, and BSPMs are readily available to purchase. The downside of the bank configuration is that if one of the electrolyzers goes out, all the others go out; and, pulling one out of the system requires that the whole system stops gas production, while the problem is being solved. Also, if one electrolyzer is performing poorly it can take the other units down in gas production. And, with common electrolyte, you have to deal with a larger amount of caustic in situations where the electrolyte has to be drained. Series connected electrolyzers act just like some Christmas tree lights, in that, when one unit goes out, the rest of them go out. Generally, series connected electrolyzers are not as efficient as parallel connected electrolyzers.

Setting up electrolyzer banks

If you use BSPMs with a common electrolyte reservoir configuration, arrange the electrolyte feed tubing and electrical connections to allow the DC current to proceed through each cell without diverting through the electrolyte. To avoid diversion of current through the electrolyte, provide a path of greater resistance through the electrolyte since the current will follow the path of least resistance. For example, for four 12 volt 10 ampere BSPMs connected in parallel-series (that is, two pairs of parallel connected panels, with the pairs connected to each other in series, to provide 4 volts, 20 amperes to each of six series connected electrolyzers, you would set up your electrical connections and electrolyte tubing connections similar to the illustrations on pages 105 and 106. You would also have to make sure that the electrolyte in the separate gas exit tubes can not connect (as shown p.106). This would be controlled by having the individual gas exit tubes sufficiently high.

Parallel connected electrolyzers

Parallel or unipolar connected electrolyzers are connected as shown at right. This type of connection can not be generally used with BSPMs because they would provide too much voltage for each electrolyzer cell. The advantage of parallel connected electrolyzers is that they are more energy efficient and the performance of an individual unit does not affect the others. If one goes down, it can be replaced or repaired while the others are still operating and producing gas. Each electrolyzer in the parallel connection has its own electrolyte reservoir. This is an advantage if the electrolyte has to be drained from the unit because you will be dealing with smaller quantities of caustic material.

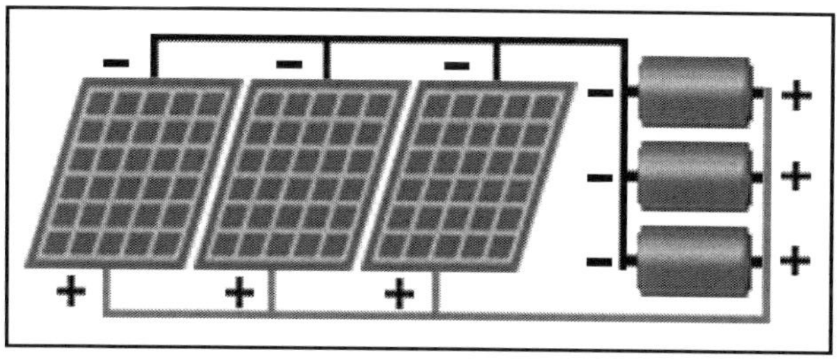

Three 4V 20 amp ESPMs connected in parallel, powering three 4V 20 amp electrolyzers connected in parallel.

A bank of parallel connected electrolyzers requires ESPMs that can deliver lower voltage and higher currents. For instance, to power 3 electrolyzers at 4 volts and 20 amps, you would have to supply 4 volts at 60 amps to power all three electrolyzers. Operating with ESPMs that produce 4 volts at 20 amps per panel, would require three panels connected in parallel to produce the 60 amps needed to power the three electrolyzers.

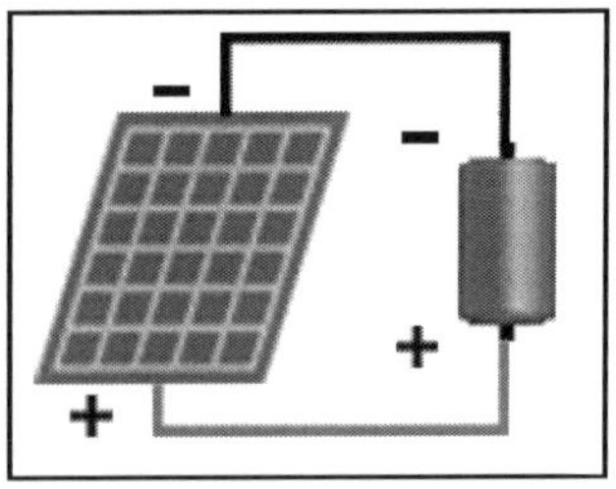

One 4V 20 amp ESPM powering one 4V 20 amp electrolyzer

Stand alone configuration

A stand alone is a configuration of one dedicated power supply for each electrolyzer. Each electrolyzer, and its power supply and water/ electrolyte reservoir is isolated from any other components in the system, except for the gas exit pipes which merge into a common storage device. This configuration requires the use of ESPMs. Thus, for an electrolyzer that you wish to input power at 4 volts at 20 amps, you would use an ESPM that produces that exact voltage and current.

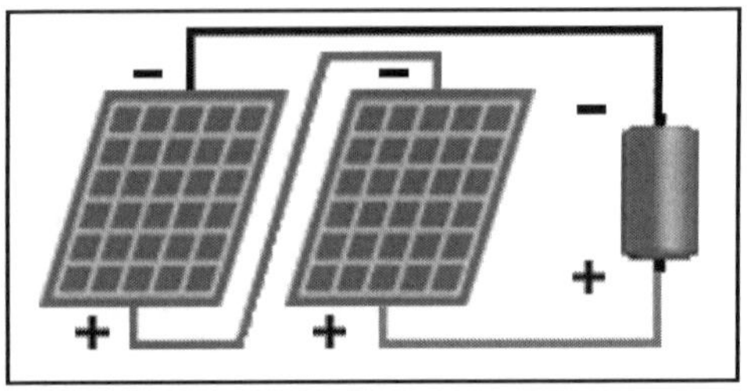

Two 2V 20 amp ESPMs connected in series, powering one 4V 20 amp electrolyzer

Large electrolyzers vs. small electrolyzer banks

In designing a solar hydrogen system, you will need to determine whether it will be more efficient and cost effective to use a bank of smaller size electrolyzers, or a larger electrolyzer or two.

It is usually more cost effective and energy efficient to use larger electrolyzers, but this is not always the case. A larger electrolyzer can easily be built by simply scaling up the dimensions of the P41.

In scaling up, it is important to increase the current capacity in the conductors and connectors. Bus and connecting wires and connectors have to be larger. Larger size conductors and connectors are necessary so that the electrical energy is not dissipated as heat. Wires and connectors can get quite hot when carrying large amounts of current. Also, since you are making hydrogen, it is wise to use amply oversized conductors. Make sure that all terminals and connectors are rated for the current you will be operating with. Similarly, you can downsize this design for micro applications and/or educational demonstrators.

Designing your own electrolyzers

It is easy enough to build an electrolyzer, but to design and build a very good one takes time and consideration. For instance, there are many types of materials that can be used for electrodes.

Some of the more exotic electrode materials are nickel coated with manganese, tungsten or ruthenium oxides for positive electrodes. These metals give quicker action for the part of the reaction that occurs at the positive electrode. Nickel plated platinum can be used on the negative electrode to increase the rate of hydrogen production. Also, gold or platinum plated nickel can be used for both electrodes, and/or plain nickel or nickel oxides. Raney type metals with their larger surface areas can also be used. Monel, as noted earlier, makes a very good electrode and is relatively inexpensive. There are also materials out there that have never been used for electrolyzer electrodes just waiting to be discovered. The shape and configuration of the electrodes and the electrolyzer vessel is another area for exploration.

Electrolyzer designs historically have emerged from particular industries. Their design parameters and requirements are quite different than what is needed to produce hydrogen from renewable energy sources. That said, there is a lot that can be learned from electrolyzer development over the years, but there is plenty of room for electrolyzer innovation, especially for those designed to operate with renewable energy power sources.

Keep in mind also, the highest performance electrolyzer may not be the best for your application. Very high performance electrolyzers are generally more expensive to make and maintain, and the materials are not as readily available. For instance, a Ferrari is a very nice high performance car, however if one simply needs to drive to the grocery store, the Ferrari would be overkill. Many people make the mistake of thinking that the highest performance electrolyzer is the best for their application. Lower cost, lower performance materials can usually do the job just as well. Reasonable costs for reasonable results is a good goal.

Electrolyzer performance testing

Alkaline electrolyzers usually operate between 1.6 volts and 2.3 volts, with the average being around 1.8 volts, and at around 20 amps current input. This is an electrical efficiency of between 54% to 78%. High pressure and high temperature (which means more expensive to maintain) electrolyzers can operate between 1.3 volts to 1.7 volts, with electrical efficiencies of 95% to 73%. However the lower voltage does not allow more current input. This means less gas production than the low pressure alkaline system, even though the process is more "electrically efficient." There is a relationship between current density and voltage requirements. More voltage is needed to drive the circuit at higher current densities.

To generally test the electrical efficiency of an electrolyzer, take a current reading and a voltage reading while the power supply is connected to the electrolyzer and it is in operation. To take a voltage reading, apply multimeter or voltmeter probes to the positive and negative electrode connectors. This reading is the voltage draw of the electrolyzer. To take the current reading in amperes, connect one probe of the multimeter or ammeter to the positive connection of the power source and the other probe to the positive lead of the electrolyzer. The current flows through the meter and the reading will give the amps being drawn at the moment.

Multiply the current reading times the voltage reading to calculate the watts being used by the electrolyzer at the moment of the reading. Divide 1.24 by the voltage reading to get the electrical efficiency of the electrolyzer at the moment of the reading. For instance, if you get a reading of 1.8 volts at 16 amps input, then the electrolyzer at the moment is operating at 28.8 watts. Divide 1.24 by 1.8 and you will see that the electrolyzer is operating at about 69% electrical efficiency.

Collect performance data for RE power sources

With most renewable energy power sources, voltage and current changes constantly due to minor or major variations of light intensity, or wind speed, etc., to name just a few variations depending on the power source. Field testing and collecting the data over time is important to get average output figures for real time situations. Average them out to figure total gas production for a season.

Testing equipment

These field tests can be monitored and information stored on a computer with inexpensive multimeters and the software that comes with them. Most electronic stores have multimeters with software available. These multimeters are quite versatile, but be sure to get one that will handle your amp output and provide current readings. You can also use data loggers to store information and then download it at periodic intervals.

For bench testing the general performance of an electrolyzer, construct a power supply controller. This will allow you to vary the current and voltage input to the electrolyzer and test its performance and characteristics. You can also measure and compare exact input in terms of current and voltage, and the output in terms of volume of gas and so on.

Measuring gas output

Gas volume, flow and pressure can be monitored in many ways. Manometers and flow indicators can be used, or the simplest method of gathering gas statistics is to invert a three gallon bucket in water in a five gallon bucket to create a floating water/gas storage tank. Although this is not as accurate as other means, it will give you an idea of how much gas is produced over a given periods of time with given voltage and current.

Gas production formulas

There are formulas to calculate gas production output under ideal conditions. These calculations are handy to make comparisons with your field and bench tests. For instance, to calculate the output of hydrogen and oxygen during an hour, first, measure the voltage at the electrolyzer terminals during operation. Then measure the amps as indicated earlier. Then, multiply the voltage times the amperage to get the power draw of the electrolyzer in watts.

For instance, if the electrolyzer draws 2 volts at 20 amps, this would be 40 watts of power being used by the electrolyzer. Next figure out how many joules are being used per second . Basically 1 watt per second equals 1 joule per second. 40 watts of power usage would be equivalent to 40 joules per second. There are 3600 seconds in an hour, so multiply 40 times 3600 seconds. This equals 144000 joules per hour (144 kj).

One liter of water yields 1,358.3 liters of hydrogen and 679.15 liters of oxygen. It takes 13,170.9 kj to disassociate one liter of water. So we divide 13,170.9 kj by 144 kj which indicates that it would take 91.46 hours to electrolyze 1 liter of water and produce 1,358.3 liters of hydrogen and 679.15 liters of oxygen.

To find out how many liters per hour of hydrogen will be produced, divide 1,358.3 liters by 91.46 hours. The result is the liters per hour that will be produced, and in this case would be 14.85 liters of hydrogen per hour.

Similarly, to find out how many liters per hour of oxygen will be produced, divide 679.15 liters by 91.46 hours. The result is the liters per hour that will be produced, in this case, 7.42 liters of oxygen per hour.

To work with cubic feet of gas rather than liters, divide by 28.317. For instance, 14.85 liters of hydrogen divided by 28.317 equals 0.524 cubic foot per hour or about one half cubic foot per hour. And, 7.42 liters of oxygen divided by 28.317 equals 0.26 cubic foot per hour. As you can see oxygen production will always be one half of the amount of hydrogen production.

Please note that these volumes are calculated for what is termed SLC (standard laboratory conditions) which is considered to be 24.47 liters at 25°C or 298°K and at a pressure which is one atmosphere or 101.3kPa.

You can also calculate for what is called STP (standard temperature and pressure). This is considered to be 22.4 liters at 0°C (273°K) and 101.3kPa (one atmosphere).

The bench tests can be correlated with field readings and the above calculations to get a rough idea of how much gas is generated during the testing period, and how much storage space will be needed for the gas within certain pressure parameters and so on.

Bench test results can be correlated with given characteristics of the microclimate for any place. Daily insolation and/or wind speed conversions to power output with given equipment parameters can be used to ballpark the general output for a given location.

For a photovoltaic system, daily insolation tables are available quite readily on the web or in various publications. For wind power there are wind tables, but exact topographical characteristics are critical to performance and general wind statistics will not usually suffice, so a site test is really necessary.

Gas Processing System

A gas processing system prepares the generated gas for end use. There are several devices involved, and a variety of possible configurations. To design an appropriate gas processing system you need to have a working knowledge of the type of gas or gases being processed. Here we are concerned with hydrogen and oxygen gas

Hydrogen history and characteristics

Hydrogen derives its name is from the Greek *hydro*, meaning water, and *genes*, meaning forming, thus "water forming." Hydrogen was recognized as a distinct substance by Cavendish in 1776 and was later given the name hydrogen by Lavoisier who noticed that water was formed when hydrogen was burned.

Hydrogen is the most abundant element in the universe. Over 90% of all the atoms, and thus about three quarters of the mass of the universe, is hydrogen. This is one very simple reason why the planet will definitely move into a hydrogen economy.

Although it is present in the atmosphere, hydrogen is not exactly a "free floater," as it is chemically very active. It combines readily with other elements and, so is locked into compounds. On this planet most hydrogen is locked into water, which make up about 70% of the earth's surface, and organic compounds. In the atmosphere it is present at only about 1 ppm (part per million).

Hydrogen is the lightest of all gases and disperses quickly if not confined. It is colorless, tasteless, odorless, and slightly soluble in water. Hydrogen can be liquified at -423°F, and can take on a metallic state under certain conditions. At about 120.7 kilajoules per gram, it has the highest energy content of any known fuel. Its atomic number is 1, its atomic symbol is H, and its atomic weight is 1.0079. Apart from the isotopes of hydrogen (protium, deuterium, and tritium), hydrogen occurs under normal conditions in two forms or kinds of molecules. These two forms are known as ortho- and para-hydrogen. They differ from one another by the spins of their electrons and nuclei.

Hydrogen can be produced by many different methods, most notably: steam reforming, electrolysis, ammonia dissociation, and partial oxidation. It can be stored for later use, as a gas, liquid or in compounds such as hydrides. It is highly flammable and explosive and can be easily ignited through static electric discharge; or by a catalyst such as platinum in air or oxygen without any other source of ignition.

Hydrogen safety

Proper precautions and safety measures recommended in the hydrogen MSDS should be followed as well as other ruling jurisdiction safety rules and guidelines when handling hydrogen. Please do a web search for the MSDS recommendations, and read them carefully! You will also need to find out what laws and regulations are in effect for yoiur location (see page 200, AHJ).

Grounding

When working with hydrogen, make sure all metal components in the system are grounded. This will bleed off any accumulating electrostatic charges that can build up on a metal surface that is insulated to the ground. If the parts are not grounded, your body will be the preferred path to ground in case of electrostatic build-up. One spark jumping from a component in the system can ignite any leaking

Grounded touch plate beside the entry to a gas processing area

hydrogen in air. In a very humid area, this is not as much of a problem, but in a dry climate, take as many precautions as possible.

Ground all ungrounded metal components no matter what the climate. Electrostatic charges accumulate on insulated conductive surfaces in many ways. Snow or sand or dust blowing across a conductor can cause static build-up. Exiting a vehicle can charge your body with static, which can be a source of ignition when approaching a hydrogen system. Never exit a vehicle and proceed to a hydrogen installation without first contacting a grounded touch plate. Whether exiting a vehicle or not, you, and others approaching the hydrogen area should always ground yourselves with a touch plate before proceeding into the area.

Hydrogen proof seals

Hydrogen easily diffuses through the smallest cracks and spaces, and it is important to perform hydrostatic tests on as many of the components and connections as possible; and to use pipe dope and tape that is compatible with hydrogen. Teflon® tape is color coded for specific applications. Yellow tape is the color code for gas line use, and would be the choice for hydrogen, although other color coded tape could be used. For pipe dope, either Lox 8® paste or Super Lox 8® is hydrogen compatible.

Restrict access to the hydrogen area

Restrict access to the experimental area to only yourself and others who understand the safety guidelines, rules and regulations that apply. The hydrogen work area should be outdoors, and/or in a structure that is fully ventilated according to guidelines. It should be a safe distance from all sources of ignition. It should also be protected from unauthorized entry. Appropriate signs should be placed according to guidelines to indicate the material and nature of the hazard area and that no source of ignition is allowed.

In other words you don't want your Uncle Bob coming over for a visit with a cigar in his mouth asking perhaps for the last time, "What's this...?" and, yes, it can happen to you – not to mention curious kids or pets and wild animals. A serious experimental station should have a cyclone fence with locked gate around the work area to keep such problems at a minimum.

Make electrical devices hydrogen safe

Any source of ignition should be strictly prohibited within your work area. Electrical connections need to be appropriate for working in a hydrogen environment and connectors should always be tightened where they will not move or be loose and cause a spark. Light and other types of switches should be rated for the hazard of the environment. Battery operated devices with switches such as flashlights can also be a hazard.

Hydrogen compared to other fuels

Because hydrogen is so light, it dissipates quite rapidly upward, and this is a good thing. Other types of fuel gases are heavier and tend to linger at ground level longer making them more dangerous. On the other hand, hydrogen has a lower flammable limit of only 4% in air and a higher flammable limit of 75%. This gives it a much wider flammable and explosive limit than other types of fuels.

Hydrogen in the presence of oxygen

Oxygen rich atmospheres also require more caution, as when working with oxygen hydrogen fuel cells. Flashback arrestors should be placed appropriately in your system so that any flame can be quenched before it can reach other parts of the system. Flashback arrestors need to be placed between the storage tank and inlet into the fuel cell, and before gas entry in the storage tank also.

Any catalytic material such as platinum, which can cause combustion of hydrogen in air or oxygen rich environments, needs to be preceded by a flashback arrestor. If you install a catalytic recombiner in the system, precede and follow the combiner with flashback arrestors.

Connections need to be secure to avoid gas leaks and loss of pressure in the system. If you are working with experimental systems such as outlined in this book, you will be changing connections and reconfiguring them more frequently, so some flexibility needs to be built into the system. With greater flexibility, comes the potential for much more leakage. Be aware of this, and adjust accordingly by adhering to all relevant precautions.

Oxygen

The name oxygen is derived from the Greek *oxys*, sharp, acid; and *genes*, meaning forming, thus "acid former." Priestley is generally credited with discovering it. Oxygen's atomic number is 8, its symbol is O, and its atomic weight is 15.9994. It is slightly soluble in water and becomes a liquid at -297°F. It has nine isotopes.

Oxygen is about 21% of the earth's atmosphere by volume, and over 49% of the earth's crust. It is colorless, odorless, and tasteless. It reacts with all elements except inert gases and it forms compounds called oxides. Although oxygen is inflammable it vigorously supports combustion with materials that are flammable.

It is used in many industries for a variety of purposes. It can be produced by electrolysis, by heating potassium chlorate with a manganese dioxide, or by fractional distillation of liquid air. It is non-toxic, and as a gas poses no hazards except for its vigorous support of combustion with flammable materials.

Oxygen and safety

Because of its support of combustion, it is important to keep oxygen separated from hydrogen both in the production of this gas in the electrolyzer, and at any other stage of gas processing and storage. Also, store oxygen away from oils and greases and other hydrocarbons. Storage of oxygen should be at least 20 feet from hydrogen tanks, or separated by a barrier at least 5 feet high and rated for a fire resistance of at least 1/2 hour.

For connections, green color coded Teflon® tape is compatible with oxygen and LOX-8®, Super LOX-8®, or Oxytite® are recommended pipe dopes.

Generated or ambient oxygen?

In your experimental system you may use the oxygen generated, or not. Hydrogen oxygen fuel cells require both gases, but hydrogen air fuel cells will not need generated oxygen because they get it from the ambient air.

Moisture and fuel cells

The end use defines the quality and quantity of the gas needed. For our purposes, the end use for the hydrogen generated is a fuel for fuel cells. Most commercial hydrogen generating units incorporate a drying mechanism that removes much of the moisture from the gas. This is not appropriate for a fuel cell system. Free water will be abundant and should be removed by a water filter or series of water filters, but it is not necessary to remove very fine aerosols by coalescer or to use water absorption techniques. The reason is that the hydrogen side of the fuel cell membrane needs to remain hydrated (moist at a certain level) to aid and maintain proton transport for the operation of the fuel cell.

Also, in terms of moisture for the hydrogen side, consider the number of stacks in the system. If there are a large number of stacks, there are more membranes to keep moist. For one stack, less moisture is needed. Design and develop water management according to your needs.

If you supply oxygen from an electrolyzer to a hydrogen-oxygen fuel cell, be sure to remove water from the oxygen. The buildup of water on the oxygen side of the fuel cell is more detrimental, because water is being formed there by the action of the fuel cell. It can drown the membrane and make it temporarily ineffective and inoperative. On the oxygen side of a gas processing system, you may wish to add an extra filter or coalescer and also a drier.

Removing contaminants

It is important to remove and/or neutralize contaminants from the gas generating system. In this particular system, potassium hydroxide is the electrolyte. It can be removed by passing the gases through a scrubber-bubbler filled with water and/or vinegar. Vinegar neutralizes the KOH. Other particulates are also picked up by the scrubber.

Scrubbers and diffusers

Liquid scrubbers remove contaminants quite effectively. They can be made even more effective by breaking up the gas into smaller bubbles with screens or diffusor/air-stones. These devices expose a greater surface area of the gas to the liquid removal agent and thus make the scrubbing more efficient. Use a larger pore diffuser for this – finer pores take more pressure and clog up faster. Better quality glass bonded silica diffusers have better resistance to KOH and do not deteriorate as quickly as other types.

Another option is to use polyethylene diffusers, and/or stainless steel or monel metal screen diffusors. There are many viable approaches. If the pores are too fine, more pressure will be needed and the diffuser will clog more easily. Our system has a regular fiberglass hardware screen, which makes a larger size bubble and thus scrubs less surface; but it does not require much pressure to push the gas through.

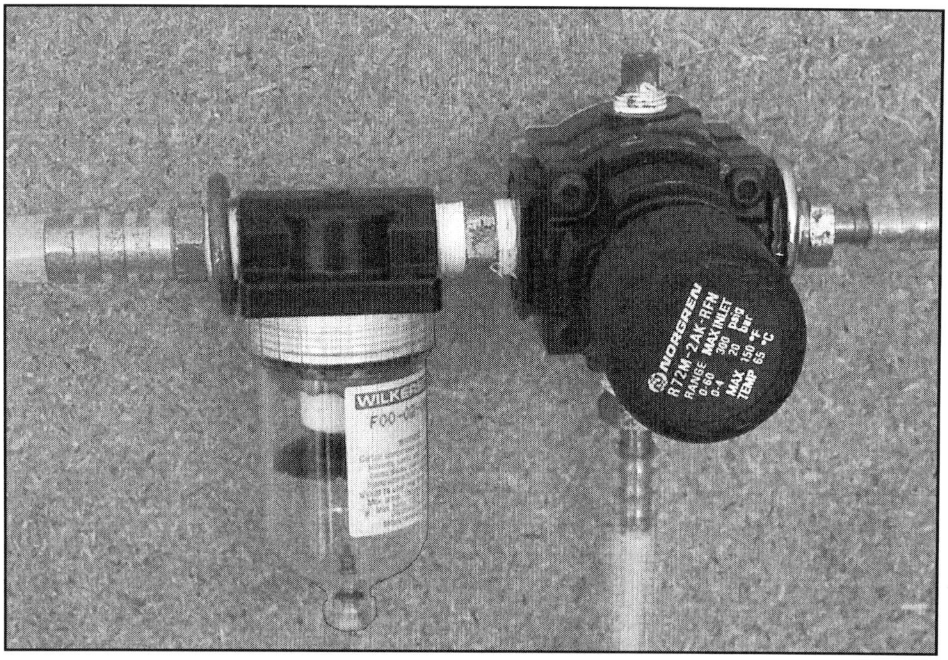

Water filter (left) and regulator (right)

Filters and coalescers

On the hydrogen side of the gas processing system there is no need to remove all the moisture because the membrane should be hydrated. So, we use one low cost water filter which does the job adequately. If you find that one filter is not sufficient, add another.

A coalescer and/or drier is not necessary, however add one if you want, based on the amount of aerosols in your particular system. There is no end use for the oxygen in our system, but if you are going to use the generated oxygen, add a coalescer to the water filter to remove as much water as possible. The number of filters and other devices, and their size/capacity will depend on how many electrolyzers there are in the system, and how large the electrolyzers are. In other words, it depends on the amount of gas to be processed.

Recombiners

The solubility of both hydrogen and oxygen in water is minimal so it does not readily diffuse and migrate. However, there is always some diffusion of gases and mixing. If you feel a need to further purify your gas, you may want a catalytic recombiner, which can be a stainless steel tube with platinum coated alumina pellets inside it. The gas is passed through the pellets, and water is formed and heat given off in the process. If you include a catalytic recombiner in your system, put a water filter on the output end and a flash arrestor on either side of the recombiner unit. There are instruction for building a catalytic recombiner on page 140.

Check valves and regulators

A check valve in the system prevents back flow of gas when the system is not operating. A regulator will provide the right amount of pressure for the fuel cell unit you are feeding. The regulator can be configured to deliver gas directly from the system, or from storage, or both. With this system, a simple valve can be used because the pressure is very low and is not enough to rupture fuel cell membranes.

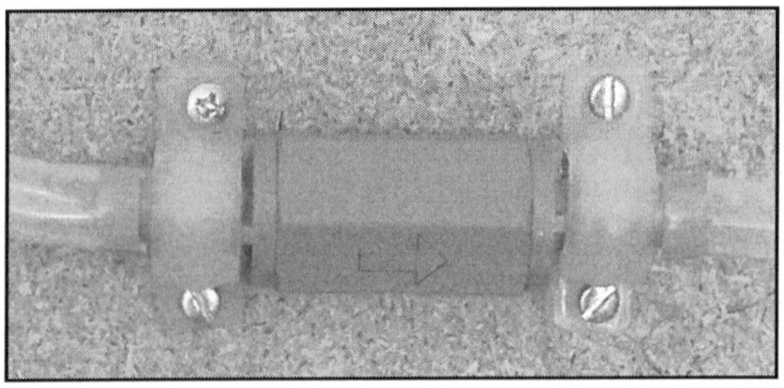

Check valve

Environment of operation

The environment in which your hydrogen and fuel cell system will operate is important to consider in the design of the components and the choice of materials used to construct them. Consider temperature. In climates where significant seasonal changes occur, contraction and expansion of components and the stress these changes will have on the system should be considered when choosing the construction technique. For instance, in one portable storage device we built, we used epoxy against PVC. It was great for one season, but when the next season rolled around, and we did our seasonal check we found that the epoxy delaminated from the surface. Cold weather contracted the materials and broke the epoxy bond. Obviously, we needed to find other options for connecting these two surfaces for cold weather applications.

Even with the best and most expensive fittings rated for hydrogen use, you can encounter problems with severe weather such as arctic conditions or desert conditions, or extremely wet conditions and/or any seasonal or daily changes – anything that creates stress or shows materials' incompatibility in rates of expansion and contraction.

System longevity

The fourth consideration is the longevity of the system. Is it experimental, with a short life expectancy, or should the system have a permanent structure? The system components in this book are used and intended for a short life experimental framework. The system is not designed for long term use with hydrogen rated components. A long service system would require more technologically mature and more expensive components, such as described in my book *Practical Hydrogen Systems*.

If your experimental design performs to your satisfaction, you can then upgrade to more durable components. High quality stainless steel components should be used in every system intended for any length of service. Other materials can be substituted, but they must stand up to the conditions required for the purpose and location of your system. Basically, most natural gas fittings, pipes and connectors will suffice, but you must do your homework and make sure they are rated for hydrogen use.

Gas processing system

The gas processing system for this particular solar hydrogen fuel cell system consists of several gas scrubbers, a gas regulator, water filter, and a few valves, check valve and flashback arrestors.

Gas scrubber

Gas scrubbers absorb contaminants from the gas that is fed through them. Liquid gas scrubbers as we describe here, provide a safety feature also, in that they act as flashback arrestors as well.

Liquid scrubbers, or bubblers as they are sometimes called, simply release the gas through a liquid. The liquid absorbs contaminants, refining the gas so that it will not be detrimental to the device being fed, in this case, a fuel cell. We chose distilled water or vinegar, and a combination of both, though there are other liquids that will work.

Vinegar and distilled water

In this gas scrubber system the liquid absorbs particulate contaminants and also removes any electrolyte aerosols that flow out of the electrolyzer with the gas. Distilled water is very effective at removing the electrolyte from the gas, but the addition of vinegar neutralizes the KOH, which enhances the scrubbing process greatly.

If you use vinegar to scrub gas, make sure it is sulfur free. Natural foods stores often have natural vinegar in bulk. This is your best bet for getting vinegar with no additives. Sulfur compounds have a negative effect on fuel cells, so be sure to avoid them. If you buy bottled vinegar, beware. Some bottles state on the front that it is natural but if you read the fine print, you will see that sulfur has been added. If you can, get red (wine) vinegar. If you use red vinegar, as the vinegar gets depleted in its neutralizing capacity, it turns a pale color which indicates that it is time to change it. White vinegar works fine also, and is also cheaper to buy, which can be important if you are going to use large quantities.

Our system has two scrubbers. Both can be filled with distilled water, or one can be filled with vinegar and the other with water. If you decide to use both vinegar and water, put the vinegar filled scrubber first in the system, followed by the water scrubber. This way the vinegar is also absorbed from the aerosol stream in the second scrubber.

If you wish, use one scrubber filled with distilled water instead of two; or you can have four scrubbers: four filled with water; or two filled with vinegar, and two with water; or one with vinegar and three with water.

Experiment with different designs

If the scrubber has a large diameter, it will hold more liquid and will not need the water or vinegar changed as often. There are many ways to construct bubblers. The main consideration is to get as much gas surface area exposed to the scrubbing fluid as possible, without creating pressure problems. The smaller the bubbles, the more surface area will be scrubbed. Also, the further the gas bubbles have to travel through the scrubbing fluid, the more they will be scrubbed. However, the height of the liquid in the scrubber adds pressure that must be overcome by the gas in the feed tube inside the scrubber, so don't make the liquid level too high.

System design largely consists of making compromises. I decided to use larger mesh screen because I wanted to use the bubblers with a number of different electrolyzers, and I did not want to restrict the gas flow unnecessarily. Also, since the bubblers would be sealed, the mesh cannot be accessed to be cleaned or replaced. With a

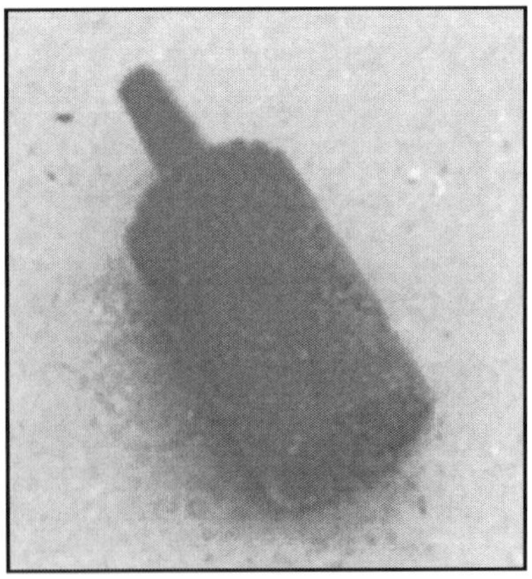

Air diffuser

finer mesh, clogging problems could occur over time.

If I used silica sand air diffusors, this clogging would definitely occur and the diffusors would have to be cleaned with muriatic acid occasionally to remove any particulate buildup

Although I do not see this as an especially serious problem, I still chose to make them as maintenance free as possible, which meant that I would have to sacrifice some scrubbing action for longevity.

I could have used a finer mesh screen, and it would have been preferable to use monel screen. However, I happened to have fiberglass screen on hand, and decided to try it. You can go down to a fairly fine mesh screen without clogging problems if you make sure that the KOH solution is as particle free as possible, that is, filtered beforehand. Of course if you design the bubblers to allow you to clean the screen, you can used a very fine mesh.

Making bubblers

Tool list

Taper pipe tap, high speed steel, bright finish 1/2"-14 NPT, McMaster-Carr, part #2525A175.

Drill bit, round shank Hss, 1/2" shank diameter, 45/64" size, 6" overall length, 3" flute length. McMaster-Carr, part #2933A57.

Value-Rite tap wrench, straight handle, style for 1/4"-1" (6-25mm) taps. McMaster-Carr, part #25605A79.

Value-Rite tap wrench, straight handle, style for 0-1/2" (1.6-12.5mm) taps. McMaster-Carr, part #25605A75.

Hacksaw, and other types of saws for metal and other materials.

Drill, either handheld or drill press.

Caulking gun.

Materials list for 2 bubblers

All should be available at local hardware or plumbing supply stores, unless otherwise noted.

1 PVC nipple, threaded on both ends, 7/8" OD, 3" length. Cut in half for two 1 1/2" pieces. Can use any other substitute that works.

Fiberglass screen, stainless mesh. Cut 2 pieces to size to cover nipple ends.

2 pieces PVC pipe 3" ID 12" length.

4 #3 caps for 3" ID PVC pipe.

6 PVDF single-barbed tube fittings, 90 degree elbow x NPT male for 3/8" tube ID, 1/2" NPT. McMaster-Carr, part #53055K191.

2 pieces 1" plastic tubing, 9 1/2" long.

Clear silicone rubber caulking.

Other parts for gas processing system

1 Filter with zinc body and polycarbonate bowl, manual drain, 1/4" pipe, 35 scfm max. McMaster-Carr part #4274K94.

1 PVC spring-loaded ball-check valve 3/8" barb x 3/8" barb, Buna-N-Seal. McMaster-Carr, part #7933K33.

2 Barb hose connectors, 3/8HB x 1/4MPT, part of reducer unit connecting from electrolyzer tube to 3 way valve, and one to fit into filter.

1 Threaded barb connector 1/2HB X 3/8MPT, barb tube reducer part, for connection from electrolyzer to 3 way valve.

Other parts for gas processing system (continued)

2 Easy-Grip PVC miniature ball valve straight, 3/8"x 3/8" Barb shut off valves for KOH supply. McMaster-Carr part #4757K18.

2 Easy-Grip PVC miniature ball valve 3-way, 3/8" barb x 3/8" barb. McMaster-Carr part #4757K58.

1 coupler, 3/8"x 1/4" FPT, for connection from electrolyzer tube to 3 way valve.

Plastic tubing, variety of sizes.

Making a bubbler

This bubbler is constructed from a 12" length of 3" inner ID (inner diameter) schedule 40 PVC pipe. Two #3 PVC caps are the bottom and top. Tube connectors are PVDF single barbed tube fitting, with a 90° elbow x NPT male for 3/8" tube ID, 1/2" NPT. Plastic 1" tubing, 9 1/2" long, extends down into the bubbler tube from the tube connector in the top cap. Fiberglass screen mounted on the lower end of the tubing breaks up the gas stream into smaller size bubbles.

Assembly

Drill and tap two holes across from each other on one of the end caps. This will be the top cap of the bubbler. The holes are the entry and exit ports for the gas, and accommodate two 90° barb connectors. Drill and tap a hole in the center of the bottom cap for the third barb connector. Trim the barb connector for the bottom cap, and one of the connectors for the top cap, so that they will be flush with the inside of the caps. One barb connector in the top cap should be left untrimmed so that tubing can be mounted on it inside the bubbler. Spread epoxy on the threads in each hole and on each connector, then screw each connector into its port. Make sure the outlets point in the right direction (see

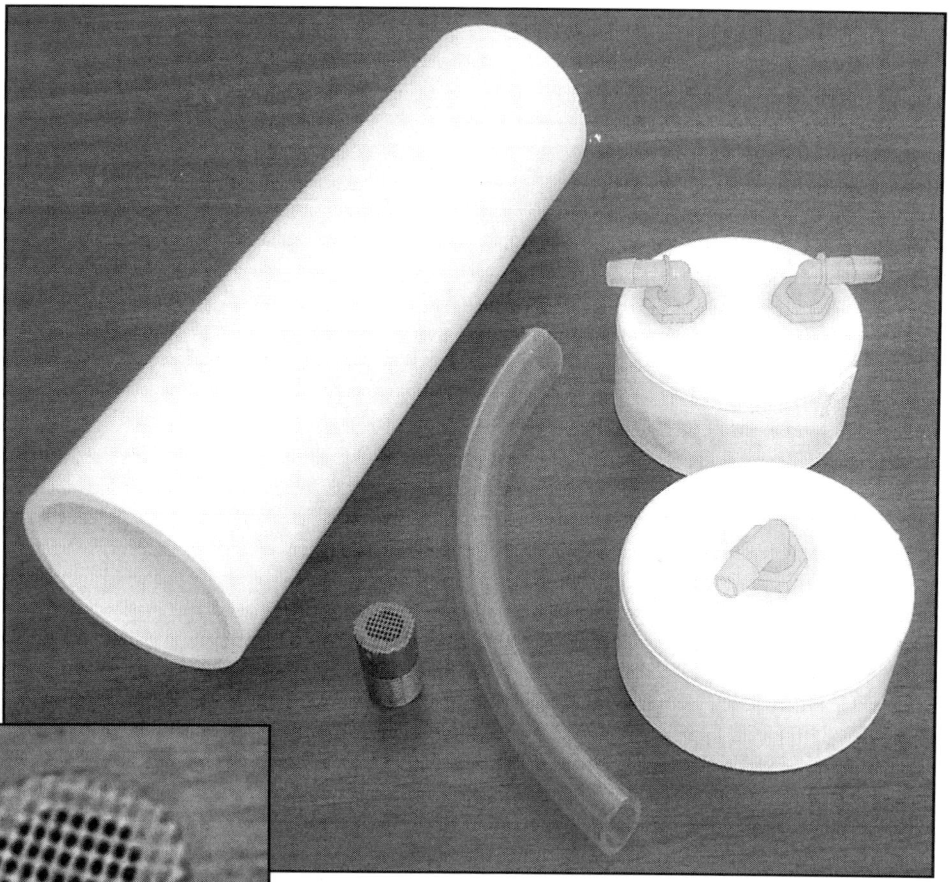

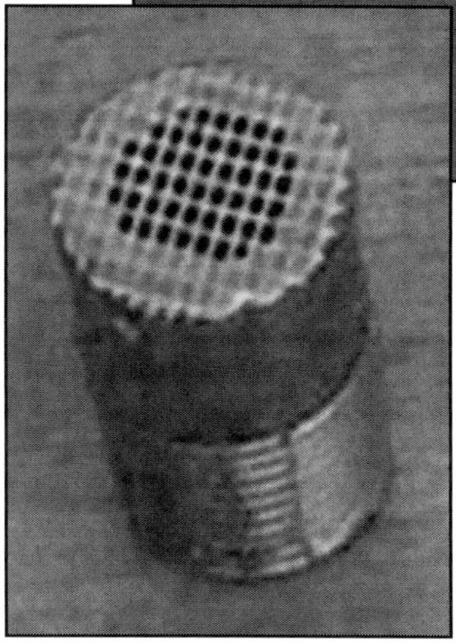

Bubbler parts, above. At left, close-up of nipple that will be mounted on the bottom of the tubing inside the bubbler.

illustration above). When they are properly seated, to ensure a good seal, add epoxy inside and outside the cap where the connectors meet the pipe cap surface. Let the epoxy dry for 24 hours.

Cut a 9 1/2" long piece of 1" plastic tubing and push one end over the piece of the barb connector that extends down from the top pipe cap. Cut a small piece of fiberglass screen in a circle the size of the outside diameter of the PVC nipple. Epoxy this to the free end of the nipple, then push the nipple into the tube. You can also apply epoxy

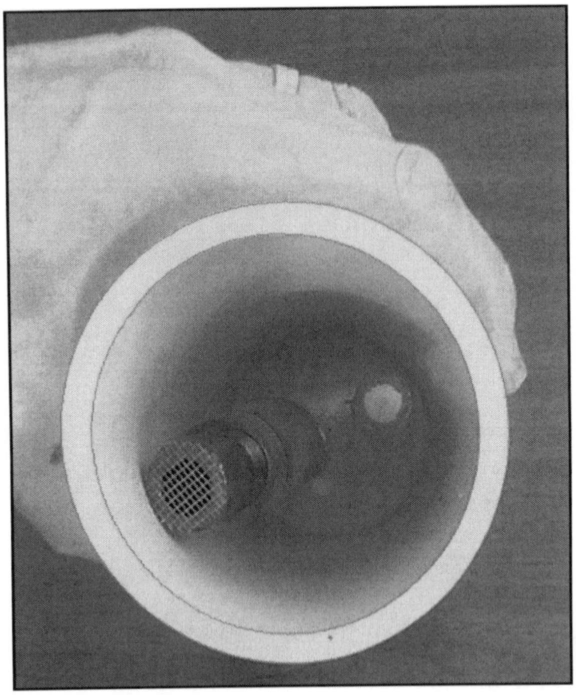

Looking up towards the top cap of the bubbler. Notice gas exit port at about two o'clock.

or silicone where the tube butts up against the inside top cap to ensure a good seal. Let it dry for 24 hours.

The bubbler operates by conducting the gas through the connector, down the tube, and into the liquid scrubber. The screen on the end of the inside tube breaks up the gas into smaller bubbles. The bubbles rise to the surface and leave through the exit port. The bottom cap connector is attached to an outside tube with a two way valve on the end. This tube is the fill/empty tube, and liquid level indicator.

Coat the inside of the top cap side walls liberally with silicone. Then coat the outside of the pipe edge with silicone for about 1 3/4" from the rim. Seat the pipe cap and press all the way down. Put a bead of silicone around the rim of the pipe and on the cap where the cap and pipe will be in contact. Smooth the silicone out with your finger or other tool to get a good, clean seal. Do the same with the bottom cap and let it dry for 24 hours.

Connecting the valves and tubing

Connect a 3 way valve to the left input port of the first bubbler with a 2 1/2" piece of 3/8" ID tubing (see photo next page). Connect another 3 way valve to the output port of the second bubbler with a 2 1/2" long piece of 3/8" ID tubing. Connect the output port of the first bubbler

Two bubblers installed in the gas processing train

to the inlet port of the second bubbler with a 5" piece of 3/8" ID tubing. Connect a 21" or 22" piece of 3/8" ID tubing to the fill/ level ports on the bottom caps of the bubblers

Connect a piece of flexible tubing about 3" long to the top outlet of the 3 way valve going to the left input port. Since the size of the tubing coming from the electrolyzer is larger than the input to the bubbler, add a reducer. I used a 3/8" x 1/4" FPT coupling, with a threaded barb connector, 1/2" HB x 3/8" MPT, and a 3/8" HB x 1/4" MPT barb. This particular combination is what was available at the local hardware store, but you may find a more suitable one piece connector reducer. This connects the 3/8" ID tubing from the top of the 3 way valve to the 1/2" ID tubing that goes to the hydrogen gas outlet on the electrolyzer.

Connect a 6 1/2" piece of 3/8" tubing to the outlet port on the second bubbler and connect that to the ball check valve. Connect a 4" piece of 3/8" tubing to the other side of the check valve. This will be connected to a 3/8"HB x 1/4" barb that has been inserted into the water filter. When inserting this connector into the water filter, use Teflon® tape and pipe dope to assure a decent seal.

Insert a 1/4"x 1/4" connector into the opposite port of the water filter, and connect it to the regulator. Use Teflon® tape and or pipe dope

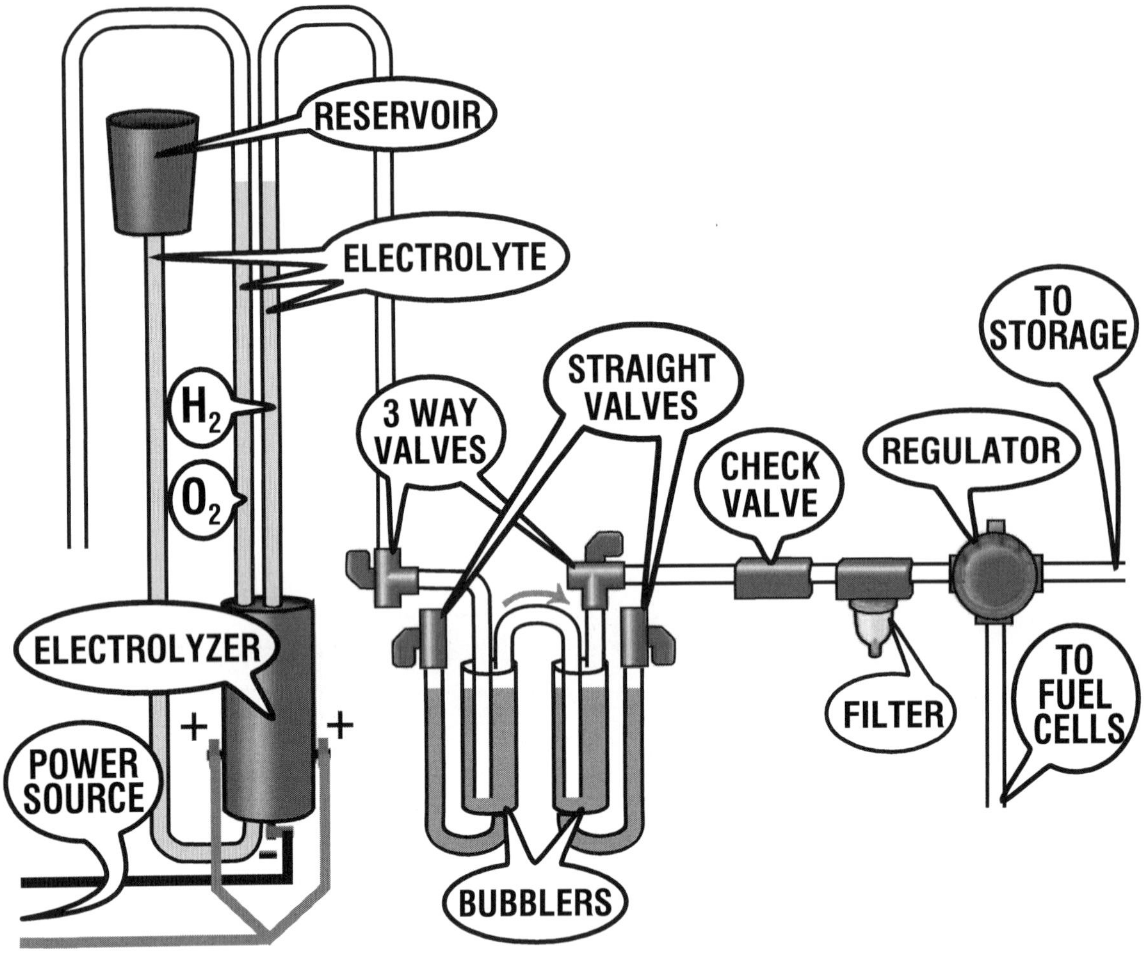

The entire gas processing system

on these threaded connections. The regulator can be configured in different ways, and you can change the configuration to suit your needs. I simply used one port for the inlet, and used two ports for output with 1/4"x 3/8" barbs, one going to the storage tank and the other going to the fuel cell stack for immediate use. I closed the 1/8" port on this regulator with a cap, however, you could insert a pressure gauge here if desired. With this regulator arrangement I can directly tap both electrolyzer and gas storage. Be sure to use Teflon® tape and pipe dope on all threaded fittings and make sure they are

The regulator and brass connectors

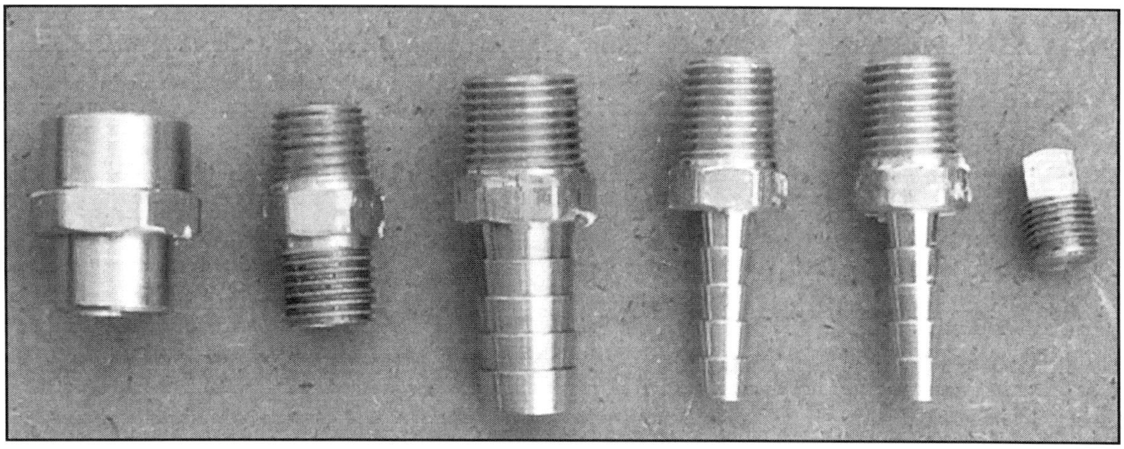

secure. Remove the bowl from the water filter and apply Teflon® tape and/or use pipe dope. If you do not do this, it will definitely leak gas, as the tolerances are not precise on the threads.

The particular filter we used for this project has a polycarbonate bowl. Polycarbonate is not well rated for use with potassium hydroxide. We decided to use this however, as the system has a scrubbing system to neutralize and absorb a good portion of the KOH solution, and the system was not intended for long service. Polycarbonate does not stand up well in harsh weather conditions with wide temperature fluctuations, and in general tends to crack quite easily.

For long service, use a filter with a stainless steel bowl. These are more expensive but necessary for a quality, trouble free system. Zinc bowls are also available, but zinc is not compatible with KOH. Filter bowls also come with either manual release or automatic release. With manual release you have to drain the bowl on a regular basis by pushing on a plunger at the bottom of the bowl. The automatic release relieves itself when it gets to a certain fullness. In this project we used a manual release, but for a long term system you might want to try an automatic release.

Mounting the system

Using adjustable pipe hangers, attach the components to a backboard. I used a 48"x 18" backboard for this purpose. Large size pipe hangers can be purchased at local hardware, plumbing supply centers or home centers. Lay out your system configuration, attach the pipe hangers to the backboard and then insert the electrolyzer, and scrubber units into the pipe hangers. Adjust for tightness so that the units are held securely. Attach the rest of the components to the backboard however you wish.

Component configuration

This system is designed to be a functioning display. Every part is in view, and the parts and components can be easily changed and the design modified during demonstrations so that the principles of the system can be understood, observed, and worked on.

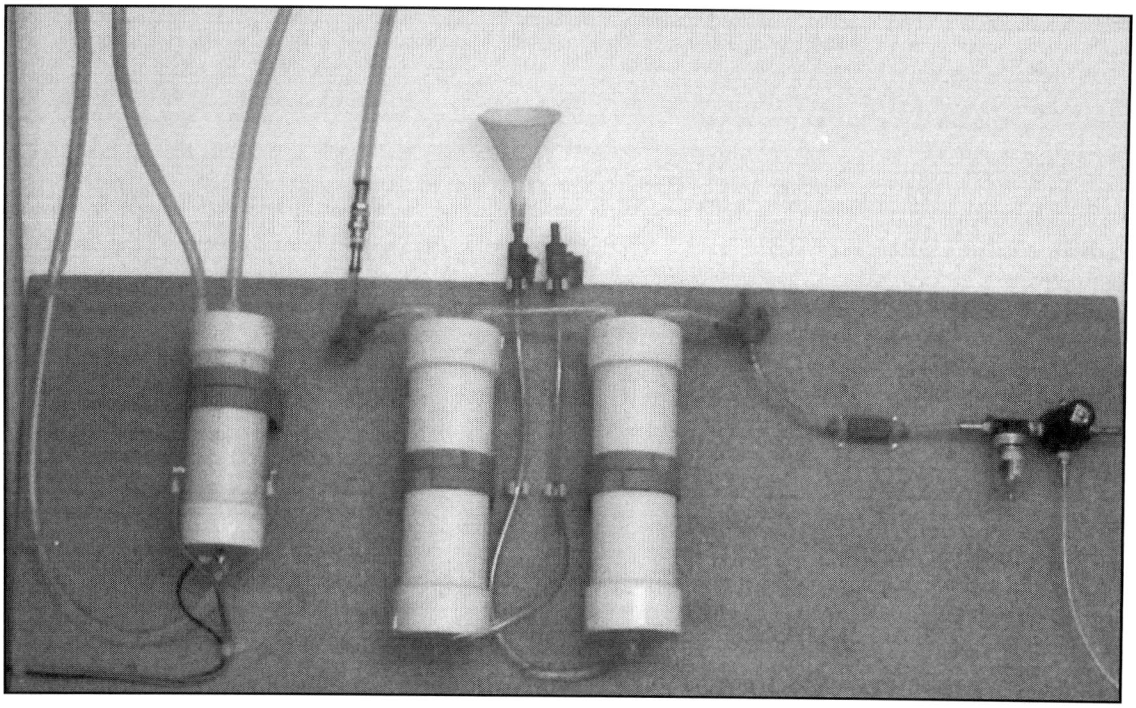

System mounted on display board

The electrolyzer and GPS system can be put together in a much smaller space. The components can be arranged to give the system a very small footprint. This particular configuration shows only the hydrogen side of the GPS.

If you intend to use the oxygen generated by this system, you can add bubblers, check valves, catalytic recombiners and whatever else you need for the oxygen side. Since this project is for a hydrogen/air fuel cell, the oxygen did not need to be processed and was shunted to a temporary storage tank without scrubbing, or filtering. A check valve was placed on the oxygen line to prevent back flow.

Additions to the system

Gas detection system

Hydrogen gas detectors located in your work area can provide warnings of any system leaks. In-line gas detectors, either hydrogen or oxygen, can provide information on the purity of the gases in the system. The most common gas detection technologies are electrochemical and catalytic.

It is important to understand the limitations, operating parameters, calibration, and maintenance for any detector that you use. Most manufacturers have plenty of information about their detectors on their websites. For instance, Neodym Systems offers low cost evaluation units, and they have a variety of configurations that allow room for experimentation.

Another safety measure is to use flame detectors in case of a hydrogen fire. Infrared detectors cannot detect hydrogen flames. You have to use an ultraviolet flame detector that has an 1800 to 2500 angstrom range. This range excludes the ultraviolet from the sun that reaches the earth, which could cause false alarms. The 1800 to 2500 angstrom range UV that comes from the sun is absorbed by the atmosphere. Hydrogen flames are very hard to see – almost invisible – and people have actually walked into hydrogen flames because they could not see them burning. False alarms can be triggered by lightening, corona discharges and welding being performed at a distance. Detectors are great if they are used properly. If they are not used properly they can be useless. Your AHJ (authority having jurisdiction, see p.200) has the information about all the necessary safety precautions you need to follow, and what they require for detection equipment.

Catalytic recombiners

You can further purify the hydrogen gas coming from the electrolyzer by adding a catalytic recombiner. This will cause oxygen that has gotten mixed in with the hydrogen to recombine with hydrogen into water molecules. The water is then drained off through a filter or coalescer, which creates a more oxygen-free hydrogen gas stream. Recombiners can be put on both the hydrogen and oxygen side of the GPS system.

Safety considerations for recombiners

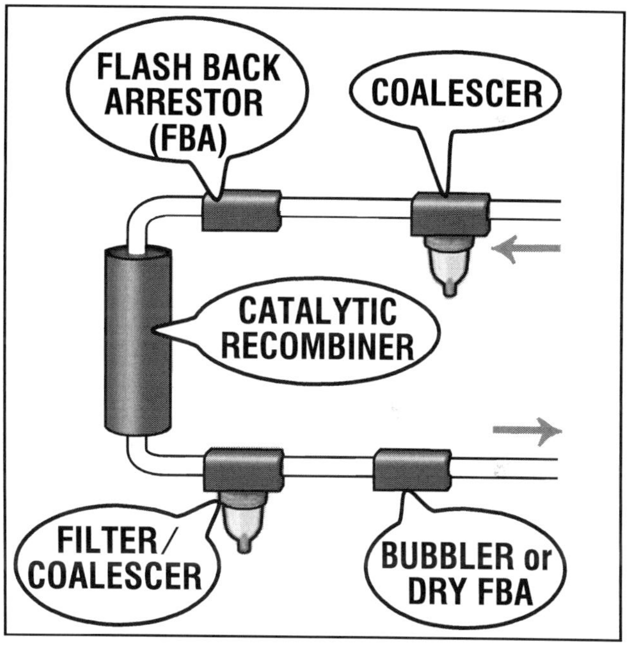

If you do add catalytic recombiners, put flashback arrestors on either side of the catalytic recombiner tubes. In the recombination process, both water and heat are created. The arrestors make it a safer process. It is also necessary to provide a drier gas to the recombiner tube. Too much moisture on the catalytic bead surfaces will impede the action.

To provide a dryer gas to the tube, precede the recombiner tube with a coalescer and then a dry type flashback arrestor, as shown at right. A liquid bubbler type arrestor would add moisture to the gas stream, which is not desirable at this point. A simple system with a coalescer followed by a dry arrestor at the feed point, and a filter/ coalescer and bubbler arrestor at the exit point is appropriate.

The use of a bubbler arrestor at the exit point humidifies the hydrogen again which makes it more suitable for fuel cell applications, as the hydrogen side of the membrane needs to be kept moist for proton conductivity. However, if you store the gas as metal hydrides, don't use a liquid bubbler on the exit end. For metal hydride storage, you want a dry flashback arrestor to keep the moisture level down. Moisture, oxygen and other contaminants will impede the hydriding action, so keep the hydrogen gas as pure and dry as possible. To use the hydrogen gas from the hydride bottle for fuel cells, you will need to humidify it by moving it through a bubbler to the fuel cell. Gas purifiers, arrestors, etc. can be purchased from companies such as Harris Calorific, Inc.

Building a recombiner

You can make your own recombiner with 0.5% platinum on 1/8" alumina pellets. These can be purchased, for instance, from Alfa Aesar (part # 89106).

A catalyst container can be constructed from a stainless steel tube with screw threads on the outside at both ends that will connect to threaded gas connectors with barbed hose connections. The stainless steel tube should be about six to eight inches long and 3/4" to 1" wide inside diameter.

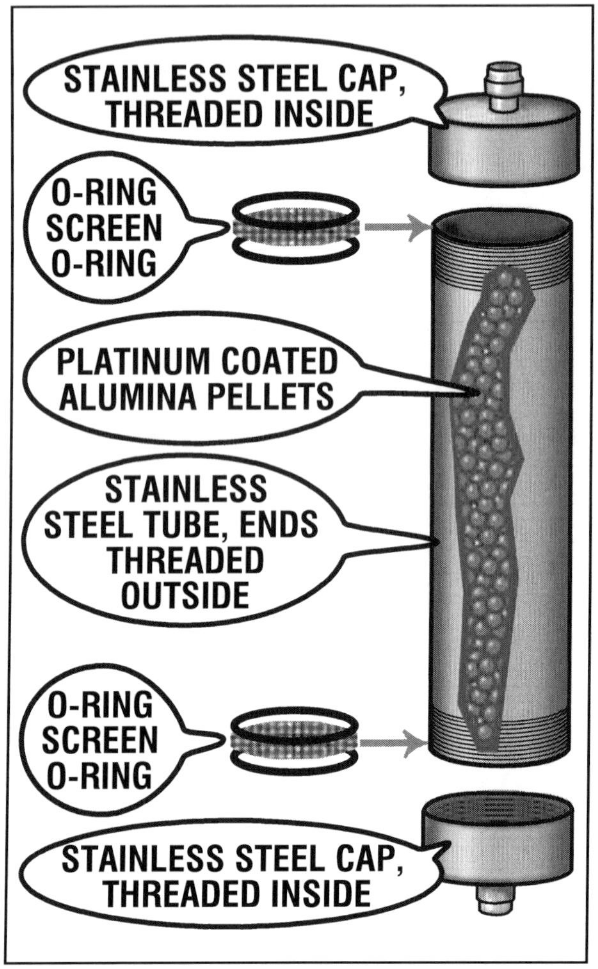

Catalytic recombiner

Four silicone or Teflon® O rings will be needed, two for the top and two for the bottom inside the tube. These hold stainless steel wire mesh that keep the catalytic beads inside the tube. The mesh should have openings smaller than 1/8" to hold in the 1/8" pellets, but the mesh openings should not be much smaller than what is necessary to retain the pellets, or the gas flow will be impeded. The recombiner should be placed in a vertical position, and the gas should enter the top and exit through the bottom.

Purging option

You can add an optional purge routine to this system by simply using bottled nitrogen with a two stage regulator.

Purging the system with an inert gas removes the air from the system before gas production. This gives a purer gas stream at the starting gate, and is an added safety measure. Without a purge system routine, you must run the system until the air is pushed out and replaced by the gases emitted by the electrolyzer.

Several experimenters have noticed that some of the experimental PEMs on the market recently are quite thin, and that if their systems or fuel cells are not purged, they get what is called in the trade "membrane blowout." The oxygen in the system air that was not purged out, and the hydrogen combine in the presence of the catalyst on the membrane, and burn micro holes in the cheaper thinner membranes, resulting in a "pop" or system failure.

I cannot recommend using MEAs with super thin membranes. It is probably best not to get anything thinner than a Nafion NRE 212. Although I have heard that some of these thinner membranes have been improved, I would be cautious about using them. You might want to experiment with a one cell configuration before investing and building stacks with these types of thinner materials. Most of these membranes have not had a great success in the market. The operating environments for fuel cells can be quite severe. Although delicate components may work in the lab under ideal conditions, they may not work well in the field. It should be noted that these membrane failures may also be the result of not sealing the MEAs appropriately.

We do not use an inert gas purging system and have had no trouble to date, but that does not mean it can't happen in the future.

In this system, you can purge with an inert gas by evacuating the system (with a hand or motorized pump) and attaching a hose from one of the stopcock valves, to a two stage regulated bottle of inert gas. Fill the entire system at about two psi. A separate purge valve can be added to the system. Bottled gas can be purchased from any gas supplier such as AirGas and Prax Air.

System purging can also be done by running the electrolyzer and immersing the tube that comes from the regulator, in a container of water. The air will be pushed out of the system and be replaced by hydrogen completely up to the point where you are releasing the gas into the water filled container. How long it takes for the hydrogen to replace the air will vary depending on how much tubing you have in the system and how fast the electrolyzer is releasing hydrogen, amongst other factors. The water filled container also acts as a flashback arrester during this process. If you are using the double 55 gallon storage tank for gas storage, at the top of the lower storage drum there is a tube that connects to the regulator hose. The bottom 55 gallon container is filled with water, and if you raise the hose to just above the water level in the tank, the water level in the tube will indicate the water level in the tank. Hold the water filled container with the regulator hose in it up to the point where the tube is slightly above the water level of the bottom drum. Immerse the tube coming from the drum into the water container, and then lower it a little bit. Whatever air was in the tube will be pushed out, and then you can connect both tubes underwater with a double barb connector.

Pressure gauges, indicators, and switches

Simple pressure gauges and or electronic pressure gauges can be included in the system for monitoring; as well as solenoid switches with transmitters to turn gas flow on and off, change flow rates, and/or just simply read pressure. If you use any electronic switch or monitor, it must conform to current safety standards for a flammable gas atmosphere. The total gas production electrolyzer and GPS system can be wired to give readouts for everything from current and voltage draw, to pressure and flow of gas. The data can be transferred to a computer for data logging via RF transmission. Infrared or visible light transmitters can also be used for this purpose.

Component upgrading

After you have experimented and are ready to finalize a system, you can upgrade to more durable hydrogen technology components. There are many types of piping materials, flexible or rigid, and many types of fittings available from a wide variety of vendors to suit your purposes. Generally, natural gas fittings will suffice for hydrogen application. Be aware however, that brass fittings are not recommended for anything more than short term experimental applications because the caustic atmosphere of KOH deteriorates brass.

Pipe dope should be rated for oxygen use on the oxygen side of the system, because some pipe dopes will deteriorate in the presence of oxygen. A simple search on the internet will help you find the appropriate product.

In certain engineering circles, pipe tape is not considered to be a sealant, but a threading aid only. This is an important consideration when dealing with hydrogen. The small size of the hydrogen atom means that it will leak through any avenue available to it. Teflon® tape can be used with natural gas fittings but the addition of pipe dope would improve the seal - or simply use pipe dope without the tape. Pipe dope (rated for the service involved) is the preferred sealing agent no matter what type of fittings you are using.

Gas storage

Hydrogen and oxygen can be stored in a variety of ways: storage bags, hollow tanks, or tanks with hydrides, or it can be stored in liquid form. There are also methods being developed such as the use of carbon nano-tubes, and metal organic frameworks.

Liquid phase storage

Liquid phase storage is compact, but it is not as economical as other forms of storage because of the energy needed to lower the temperature of the gases. It also requires a doubled wall Dewar type of container to hold the liquid hydrogen and oxygen. However, recent advances in magnetic refrigeration may, in the near future, make it viable to develop a renewable energy liquification system for hydrogen. Large capacity Dewar containers are available on the surplus market for cryogenic storage of gases at relatively reasonable prices.

The advantages of magnetic refrigeration are that it does not use environmentally damaging chemicals, and, at present it is about 60% efficient, compared to 40% efficiency for currently available gas compression refrigeration units that rely on a vapor cycle.

Hydride storage

Hydrides contain a relatively large quantity of gas in a small area. This technology is compatible with renewable energy systems, but the drawback is the amount of energy necessary to maintain such a storage system. Dedicated photovoltaic panels or wind generators would need to be added to compress the gas. Hydride storage has a lot of promise for compact hydrogen storage, and is a good avenue of inquiry for the renewable energy enthusiast. An explanation of the hydride process and how to work with hydrides is discussed in my book *Practical Hydrogen Systems: an Experimenter's Guide.*

Metal organic frameworks

Currently, metal organic frameworks (MOFs) also look very promising for solar hydrogen systems. These materials do not require high pressures and high temperatures to operate the system, so they do not require the level of extra energy input that hydride systems need. With only modest pressure at room temperatures, hydrogen can be adsorbed and retrieved in MOFs. One gram of a metal organic framework can have a surface area of 3000 square meters. It is also a very inexpensive material.

Low tech alternatives

Probably the most convenient gas storage options for the average renewable energy experimenter are low pressure storage in bags; and low, medium or high pressure storage in hollow tanks.

Containers for low pressure storage (under about 60 psi) can be bladder tanks, pillow tanks, collapsible pressure vessels or LPG tanks. Most flexible tanks store gases at just a little above atmospheric pressure. Manufacturers such as Aero Tec Laboratories offer flexible storage devices.

LPG tanks or other tanks made with a good grade of stainless steel are sufficient for medium pressure storage (60 to 125 psi). Stainless steel with a high nickel content and very low carbon content is more resistant to hydriding, otherwise known as embrittlement. (Hydriding is a great characteristic for the storage medium in a hydride storage system, but it's not a good characteristic for containment tanks!)

Embrittlement occurs when the metal absorbs hydrogen which can cause the metal to crack. If you make your own tanks, to prevent hydrogen from getting into the metal, do not use water while welding. Also, preheat the metal before welding and allow the weld to cool slowly. Embrittlement proceeds faster at high pressures and higher temperatures.

High pressure storage, around 2000 psi and up requires specially constructed high pressure cylinders.

Double drum storage

Short term storage can also be accomplished in tanks or drums that would not be considered or used for long term storage. For instance, I use recycled 55 gallon drums for oxygen and hydrogen storage. These drums are very inexpensive and I can quickly weld the tops on, fabricate a relief port and inlet/outlet port quite easily. With some drums I have epoxied the tops on. For me it works, as I do not use them for more than one season. The drums are readily available and the price is right.

I do have to be concerned about what was stored in the drums beforehand. Contaminating the hydrogen supply with prior drum contents can be a problem. The drums must be absolutely clean from contaminating substances. Although I use this temporary storage method for my experiments, I must caution the reader that storage in

Two recycled 55 gallon drums for short term gas storage

any thing other than your AHJ approved storage devices is not an officially sanctioned practice and is not considered safe.

For drum storage, two 55 gallon drums are needed. One drum is securely positioned above the other. The top drum has a constantly open vent which releases any excess gas that cannot be stored in the drums. The bottom drum has a gas inlet near the top of the drum. It has an outlet on the other side near the bottom. A transfer tube connects the bottom tank to the top tank. The bottom tank is filled with water and when gas is routed to the tank, it pushes the water through the tube and through to the top tank. The gas is partially pressurized by the weight of the water in the top tank. It is a simple affair and works quite well.

Any type of general gas or barb fittings can be used to make the gas outlet port and inlet and transfer ports. I usually epoxy these into place. The regulator needs to be positioned at a height above where the gas feeds in to the bottom drum, so that water does not flow into the regulator when the bottom drum is filled with water. I could expand my storage by connecting up more pairs of drums.

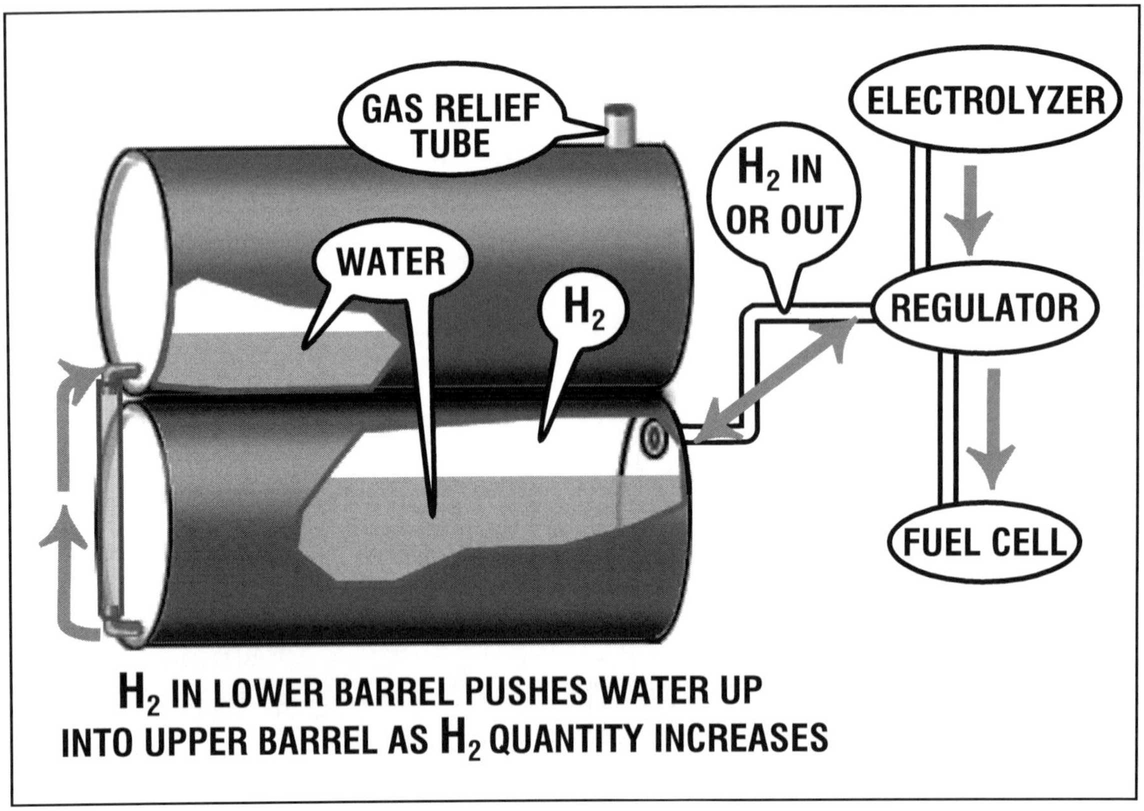

Low pressure storage in double tanks using water pressure

Floating tank storage

A water tank storage device can be built, very much like those used to store methane in some rural areas. This consists of an inverted drum or tank within another tank or drum that is filled with water. As the gas fills the inner tank, the inner tank floats upwards in the water, and is kept from tilting by guide rails or guy wires. A stop keeps the tank from rising totally out of the water. For a more permanent setup, round concrete spring tiles can be put in the ground and a inverted tank inserted for larger area gas storage. This is an easily built structure and relatively inexpensive, the most costly part usually being the fabrication of the inverted tank. Tanks can be made from fiberglass, and other resin cloth type products. Large size tanks can also be purchased from such places as Aquatic Eco Systems, or other agricultural suppliers.

Calculating tank capacity

To calculate the cubic foot capacity of a cylindrical tank, multiply 0.79 times the diameter. Multiply this by the diameter again and then multiply that result by the length. For instance, for a drum or container 7 feet in diameter and 10.4 feet in length, your calculation and result would be: 0.79 x 7 x 7 x 10.4 = 402.584 cubic feet. To convert this to US gallons, multiply 7.5 x 402.584 (there are roughly 7.5 US gallons per cubic foot). So the total gallon capacity for this tank is about 3019.38 US gallons.

The above equation gives a ballpark figure for storage at atmospheric pressure (14.75 psi). If you are considering a higher pressure storage system, such as around 60 psi, you could store about four times as much gas in the above example in the same size tank.

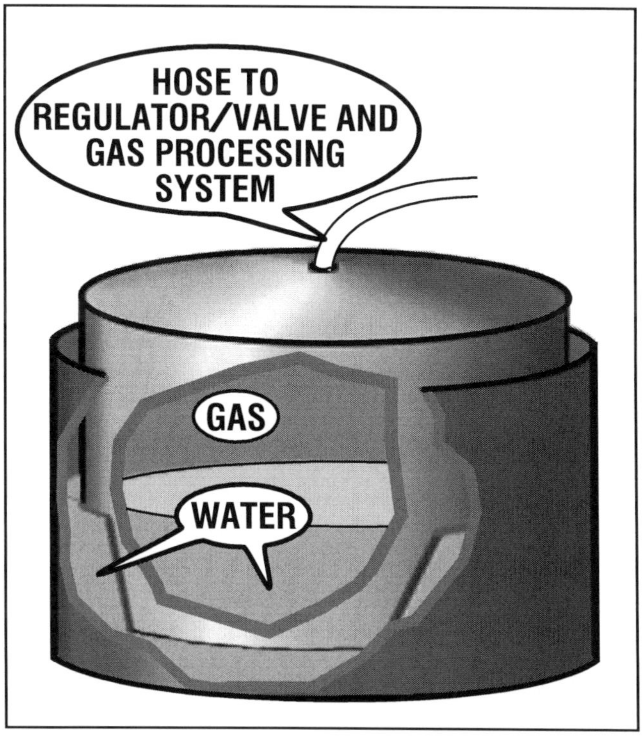

Floating tank storage

Adding pressure

If you want more pressure for storage than your pressured system supplies, use a compressor such as a diaphragm type that is rated for hydrogen use. Be sure that tanks, hoses and fittings are rated for the pressure desired. Compressors of this type can be purchased from such suppliers as KNF Neuberger and others. These diaphragm pumps keep gases contaminant and oil free, and usually have Teflon® coated diaphragms and ryton heads, which allow them to operate properly in the caustic atmosphere of aerosol potassium hydroxide solution that is used in the electrolyzer for this system.

A more detailed discussion of pressure and hydrogen systems can be found in *Practical Hydrogen Systems: an Experimenter's Guide.*

Safe storage

Storage of hydrogen should be in a protected area, preferably fenced off, with signs indicating the presence of flammable hydrogen, and signs saying that no smoking or ignition devices are allowed in the area. Drums and tanks should be grounded to dissipate static electricity and carry off induced charges from nearby lighting strikes.

Setup and check the system

1. Make sure all tubing is connected in the system and is secure. This includes all tubing related to the electrolyzer as well as the gas processing and storage components.
2. Open the valves on the bubblers and fill with vinegar and/or water to the top of the bubblers. Close the valves after filling.
3. Check to see if the valves are turned appropriately to direct the flow of gases where you want them to go. If you are going to purge, do so, and then reset the valves for start up.
4. Fill the reservoir with electrolyte and make sure the electrical connections are made and are secure to the solar power supply.

5. Turn on the power to the electrolyzer and be sure the regulator is open to the storage and other lines.
6. Observe each component while operating to see if any leaks or problems occur.

If everything checks out, you are ready to connect to the fuel cell stack.

Planar Fuel Cell Stack

Fuel cell basics

Simply put, a fuel cell is an energy conversion device. It has no moving parts and thus operates silently. In the fuel cell process, energy is released as heat and electricity.

The process is: hydrogen is fed to one catalyst electrode, which facilitates the separation of the hydrogen atoms into electrons and protons. The protons or hydrogen ions move through the membrane toward the other catalyst, which is fed with oxygen. The stripped electrons cannot pass through the solid electrolyte membrane or liquid electrolyte, so they must be routed through an external circuit. The external circuit contains an electrical load such as a motor or light bulb, etc., and leads to the other catalytic electrode, where the protons and electrons recombine and bond with oxygen to create water molecules.

If you would like to read a thorough treatment of the history and basic functioning of fuel cells, refer to my book *Build Your Own Fuel Cells.*

Types of fuel cells

There are many types of fuel cells. The most common types are:

AFC	alkaline electrolyte potassium hydroxide
PEMFC	(proton exchange membrane) uses fluoropolymer or similar type membranes such as SPEEK. Microbial and direct methanol fuel cells fall into this category also.
PAFC	electrolyte phosphoric acid
MCFC	electrolyte molten carbonate
SOFC	electrolyte solid oxide

Each of these fuel cells is named or defined by the electrolyte used in it, that is alkaline, phosphoric acid, molten carbonate, solid oxide, proton exchange membrane, etc.

The only deviation from this pattern is the DMFC (direct methanol fuel cell) which uses methanol as a fuel without intermediate reforming; and microbial fuel cells that use sugar as a fuel and derive current from the metabolic activity of yeast. Both types use a solid ion exchange membrane type electrolyte (proton exchange membrane).

PEMFCs have a solid ion exchange membrane made of sulfonated fluoropolymer, or a sulfonated polyetheretherketone (SPEEK) which is the electrolyte, and for the most part uses platinum catalysts. There are other materials in use for combination type membranes. It should be mentioned that currently SPEEK membranes do not hold up as well as the fluoropolymers such as Nafion®, but research is ongoing to produce a more reliable and longer lasting membrane.

At the present time, the Nogoya Institute in Japan is making progress developing a new glass based electrolyte that is much less expensive than fluoropolymer membranes, but just as durable. In the near future, these membranes may replace Nafion® and SPEEK in PEM cells.

SOFCs have solid yttria stabilized zirconia as an electrolyte and perovskites as a catalyst.

PAFCs have a liquid phosphoric acid as an electrolyte, and platinum catalysts.

MCFCs have a liquid alkali carbonate mixture and nickel catalysts.

AFCs have a liquid potassium hydroxide catalyst and platinum electrodes.

All of these fuel cells have different operating temperatures. For instance, alkaline cells run from 50°-200°C, PEM cells run from 50°-100°C, phosphoric acid run at about 220°C, molten carbonate run at about 650°C, and solid oxide run at about 500°-1000°C.

PEM fuel cell configurations

For the purpose of this book we decided to work with the proton exchange membrane cell. Proton exchange membrane or PEM cells are easy to work with, and come in a variety of shapes.

The most common configuration is the block type fuel cell stack. Other configurations are tubular and planar (flat) stacks.

Different shapes serve different purposes. Block stacks are convenient to use in some applications and planar stacks are easier in others For instance, a laptop computer would do well to use a planar configuration rather than a block configuration. Generally, fuel cells can come in any imaginable shape and size, and be designed specifically for a wide variety of applications.

Planar fuel cell stacks

We decided to work with a planar design in order to compare the efficiency of block stacks to planar stacks.

Block stacks, although convenient in some respects, have problems with water retention and drainage. This can be overcome, but it is never the less a complication. Block stacks can also require fans to force pull air through larger stacks, and this adds a rather dumb note to energy efficiency.

Our planar design does not rely on convection/air or pure oxygen feed (see *Build Your Own Fuel Cells* for more about convection and oxygen/hydrogen fuel cells). The planar design simply relies on ambient air ports. The intent was to see if these ports would give greater air intake, as well as more efficient water dispersion. We surmised that this design would take care of these two problems, and reduce maintenance problems (and thus cost) over a period of time.

As with the other components of this project, primary considerations were cost, availability of parts, and reasonable construction methods that do not require high tech tools.

The result of this endeavor was the L79 planar stack. Although we are still experimenting and tweaking the design, we believe it is a good design to present with the rest of the hydrogen system.

If you have never built or tinkered with fuel cells before, I suggest that you first work with a more conventional fuel cell design, before attempting to build a planar stack like the L79. My book *Build Your Own Fuel Cells* will give you a good background in fuel cell design and construction so that you will be better able to construct and/or design a planar stack.

Build the L79 planar fuel cell stack

Tools list

Available from local hardware stores, unless otherwise noted.

- Scotch Brite® Pads
- Tray, tongs
- Sand paper or fine file
- Caulking gun
- 1/8" router bit, Micro-Mark, part #60719
- Drill press or milling machine
- Multimeter (electronics store)
- Laser printer, local print shop or copy center
- Drill, either handheld or drill press
- Hacksaw, and other types of saws to cut metal and other items

Materials list

All should be available from local hardware stores, unless otherwise noted.

Press-n-Peel PCB transfer film, Techniks, Inc., #20PNPB, pack of 20 sheets, PNP blue.

Clear silicone rubber caulking

Plating kits and supplies, Caswell Plating

Barb hose connectors, one 1/4", and one 1/2"

6 Screws, round head machine, stainless steel, 6-32x 1 1/2" ;
6 washers, 7/32", stainless steel;
6 nuts, wing or otherwise, 6-32(1/8) USS, stainless steel

3 Aluminum channel bars, 12" long, 3/8"x 3/8".

Silicone rubber gasket, .020" thick, one 12"x 12" sheet, McMaster-Carr part #86435K45.

12 Membrane electrode assemblies, see page 201 for suppliers.

1 sheet PVC, 12"x 12", 1/2" thick, McMaster-Carr, part #8747K116.

2 Copper clad PC boards, one sided, 8"x 10", 1/16" thick. Circuit Specialists part# 22-264; or other electronics/PC board supplier. Make sure the dimensions are correct. Larger boards can be cut to size.

1 Silicone rubber sheet, 3/32" thick, 12"x 12", 40A durometer plain back. McMaster-Carr part #8632K43

1# Ferric chloride etchant. Circuit Specialists.

Constructing the L79

The L79 is easy to build and uses a few common materials in a unique way (see illustration, next page). It consists of a six layered sandwich composed of one PVC end plate, two PC (printed circuit) board electrode/gas flow field plates, one 12 PEM MEA (membrane electrode assembly) layer, and 2 rubber gaskets, one of which also acts as a gas supply line.

This design allows single stack units to be racked next to each other and fed by a gas manifold. The design can easily be altered so that a wider area stack can be constructed with more PEM cells on it, or with PEM cells that have a larger surface area for more current output.

Each cell in this planar stack is connected in series, that is, the positive electrode of one cell is connected to the negative of the next cell and so on. In a block stack this is accomplished by using bipolar plates, but in the L79 we use circuit board traces with tab wire connects to perform the same function as the bipolar plates. The L79 can also be built in a parallel configuration or a series-parallel configuration depending on the current and voltage desired.

Selecting materials for the electrode/gas flow field

In the L79 design, both the gas flow fields and electrodes are made from single sided copper clad circuit board. Copper clad circuit board comes in a variety of sizes and is either clad with copper on one side only, or both sides. Boards also differ in the type of base material they are made of. This fuel cell requires the FR-4 glass epoxy resin base, clad with copper on one side only. We used a 1 ounce, which is coated with copper to a thickness of .0014".

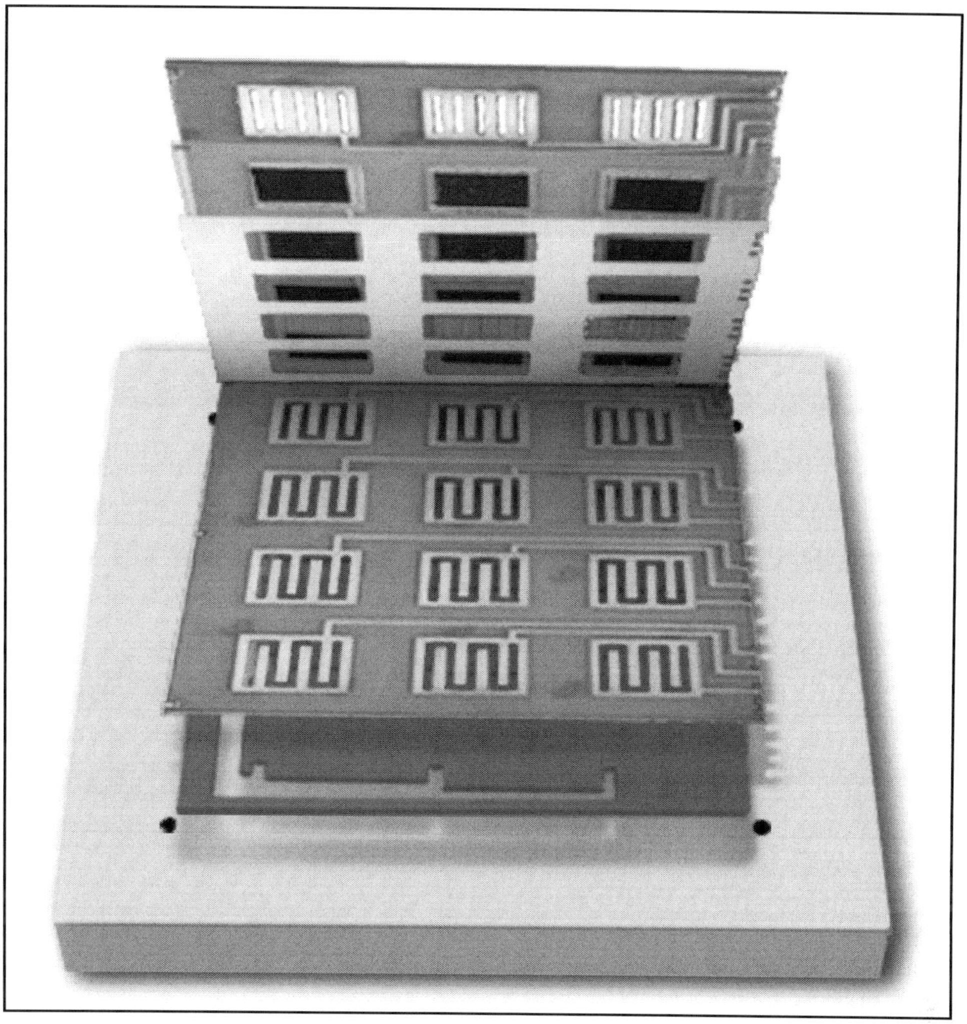

The component layers of the L79 fuel cell stack

The FR-4 was chosen because it is very stiff, which is necessary for this design. It is important for the electrode grid structure to be as stiff as possible so that it maintains an even contact surface over the membrane surfaces within the assembly, especially when it is tightened with the pressure adjustment screws.

You can experiment with other types of copper clad board with different base compositions, just remember that stiffness and machinability are the key concerns.

Preparing the electrodes for template transfer

The first step in fabricating the PC board electrodes is to clean each board thoroughly. You can use a Scotch Brite® pad and water to clean the circuit board. Even though the circuit board looks shiny and very clean, rub it all over with the pad and then rinse with water.

After you clean the board, be sure not to touch the copper surface with your fingers as this will leave minute traces of oils and dirt that can interfere with the processes of image transfer, plating, and the electrical performance of the surface. A good habit for working with copper clad is to wear clean thin cotton gloves to protect the surface of the board while you handle it.

Dry the surface of the board with a lint free cloth. Inspect the edges of the circuit board for burrs. If there are any, file them off. The idea is to get a clean smooth surface, so that the transfer film that will lie flat on the copper surface. Be sure that after you clean and wash the board that no residue is left on the surface.

Transferring the templates

The templates for the oxygen and hydrogen flow fields/ electrodes in this book (pages 203-204) are intended to be printed on plastic transfer film and then transferred to copper circuit board for etching. This method is commonly used to transfer printed circuit board designs to the copper surface of the boards. We used Press-n-Peel®, which is a plastic film that has an emulsion on one side and comes in sheets of 8½"x 11" size. The plastic sheet is loaded into the printer just as any other 8½"x 11" paper would be.

Use a laser printer to print the templates for the two circuit boards on the film. Print the image on the dull (emulsion) side of the transfer film. If you do not have a laser printer, you can print out a paper copy on whatever type of printer you have and then get a laser copy at a copy shop.

If you use a laser printer, do not let it warm up before printing. Print the template to the transfer paper immediately after turning on the laser printer. The reason is that the heat from a laser printer will affect the surface of the transfer film and can produce defects in the image. It is important to get a very sharp, clean, well covered, distinct image, as the transfer film will pick up the slightest imperfections.

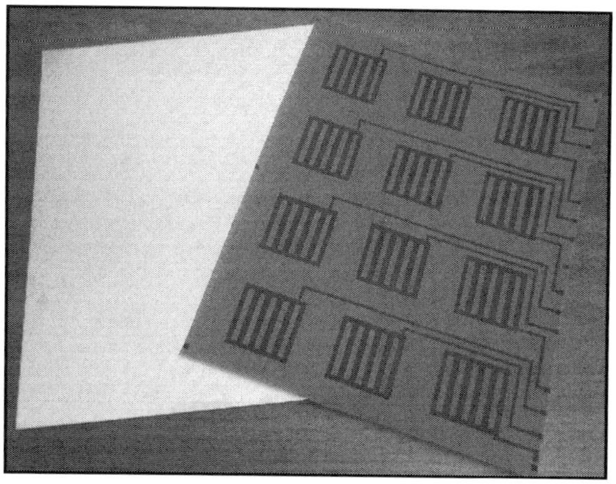

There are two circuit board templates, the oxygen side template, and the hydrogen side template. When you

Above, the oxygen electrode template printed on transfer film. Below, lay the film emulsion side down on the circuit board, and secure the film to the circuit board.

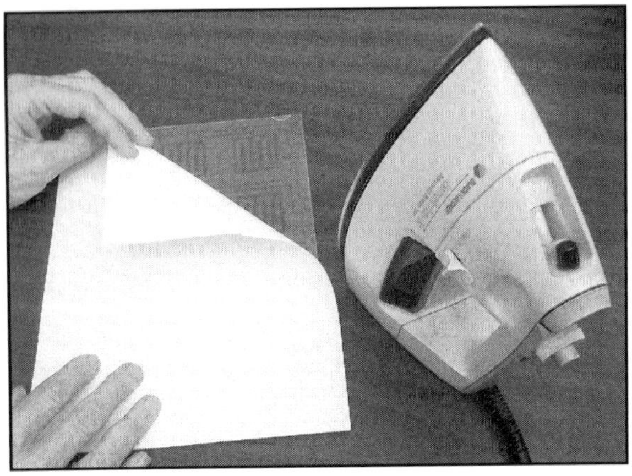

Use a piece of paper between the iron and the transfer film

have finished printing your templates on the transfer film, lay the film emulsion side down on the copper surface of each of the circuit boards.

For Press-n-Peel®, heat from a clothes iron is applied to transfer the template to the copper. The suggested starting temperature is 275° to 325° F, which is usually the acrylic and polyester setting. Not all irons are the same, and you may have to experiment a little to get the right heat. I found that using the higher "linen" setting worked better for these large sized boards.

You can attach the transfer film to the board with some adhesive dots as I did to hold the film in place on the edges when you start to iron. Place a piece of paper between the iron and the transfer film to help the iron glide better over the surface. Move the iron very slowly over the surface area of the whole board making sure to not miss the edges.

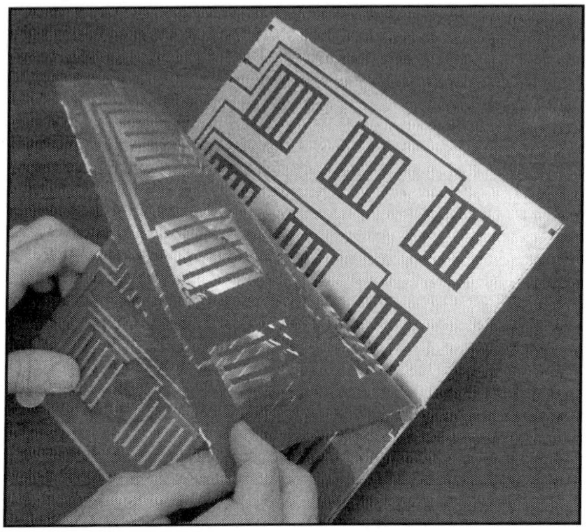

When ironing is complete, peel back the transfer film.

After about ten minutes of slowly moving the iron around on the surface, lift a bit of the film up from the surface of the copper clad and see if the image sticks to the surface of the copper. If it is not sticking, or parts are missing, iron over those particular parts or go over the surface again slowly until the whole image is transferred.

Keep checking the progress of the transfer by gently peeling the film from the copper surface every once and a while to see if the image is totally and satisfactorily transferred.

Please note that there must not be any continuity breaks in the copper surface. This does not mean no blemishes or imperfections, but does mean that if you put a multimeter probe on one side of a copper trace there should be a continuity beep when the probe is placed on the other side of the copper trace. Take your time with this step. Keep ironing until all of the transfer is complete. This can be a tedious operation but it is a very important to do it well.

When the ironing is completed to your satisfaction, take the circuit board and transfer sheet, and run it under cold water. After it has cooled down, slowly peel the transfer sheet off. You should see a perfect representation of the electrode grids in a blue color covering the parts that will remain copper clad after etching. Full directions come with each set of transfer sheets.

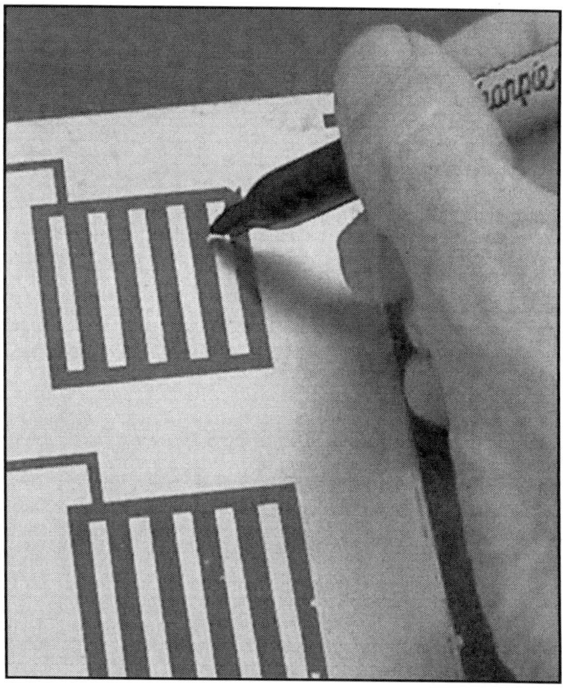

Small imperfections can be corrected with a marking pen

If there are minor imperfections in the transfer ink representations of the electrodes, you can make corrections by filling in small missed spots with a fine tipped Sharpie® marking pen.

Etching the board

You will need a container to etch the board in. This can be anything from a photo developing tray to a Rubbermaid® container. Do not use a metal container, spoon or other utensil made of metal. It must be plastic or glass. Do not use any containers that will later be used for food.

A small cat litter box works well and can be purchased from any hardware or pet store for a few dollars. Get a size that will easily accommodate the size of the circuit board with a little wiggle room, but not too much bigger, as that would require more etchant – which would be a waste just to accommodate the oversized container.

When working with etchant solution, wear rubber gloves to keep the chemicals from contacting and staining your skin. Also, wear old clothes. The stain from the ferric chloride will ruin any clothing and the stain will not come out. Make sure you have plenty of rags around to wipe up spills, and follow all the safety recommendations on the package. Always pour the ferric chloride dry etchant powder into the water. Do not pour the water onto the ferric chloride. When you pour the ferric chloride into the water, do so slowly. Use distilled water for this solution.

About 1/2# of ferric chloride at most is needed to etch both boards; 1/4# is supposed to be enough to etch 200 square inches. Each board is about 80 square inches (8"x 10"), but more etchant may be required. Follow the directions that come with etchant for mixing. One pound of etchant requires 60 ounces of water, so for one or two boards, you need 15 ounces of distilled water for each 1/4# (4 ounces) of etchant powder. Mix the crystals in with the water and fill the tub enough so that the solution will cover the top of the circuit board during the etching process.

Use only distilled and very warm to slightly hot water to mix with the etchant powder. A warm etching solution works much faster than a cold solution, so make just enough solution at one time to etch one board. The solution cools as it sits during the etching process and thus the etching process slows down. A heating pad turned on high placed under the etching tray can help to maintain a good etching temperature.

Bathing the plates in etchant solution

To etch, slip the board into the tub filled with solution. Every few minutes, agitate the board by moving it around or tilting the tray back and forth to move the solution across it. This motion is necessary and makes the etching go much faster. As you do this, you can see the copper disappearing in the areas that are not covered by the resist.

When all the copper has been removed where it is supposed to be removed, take the board out of the solution and rinse with water. This should leave a perfect image of the electrode grids in blue. Do not leave the board in the etchant longer than necessary, or the solution will begin to undercut the electrode pattern. This is not a process in which you can go away for a while and then come back. You must monitor the whole process, agitate the tray and immediately remove each board when the process is complete.

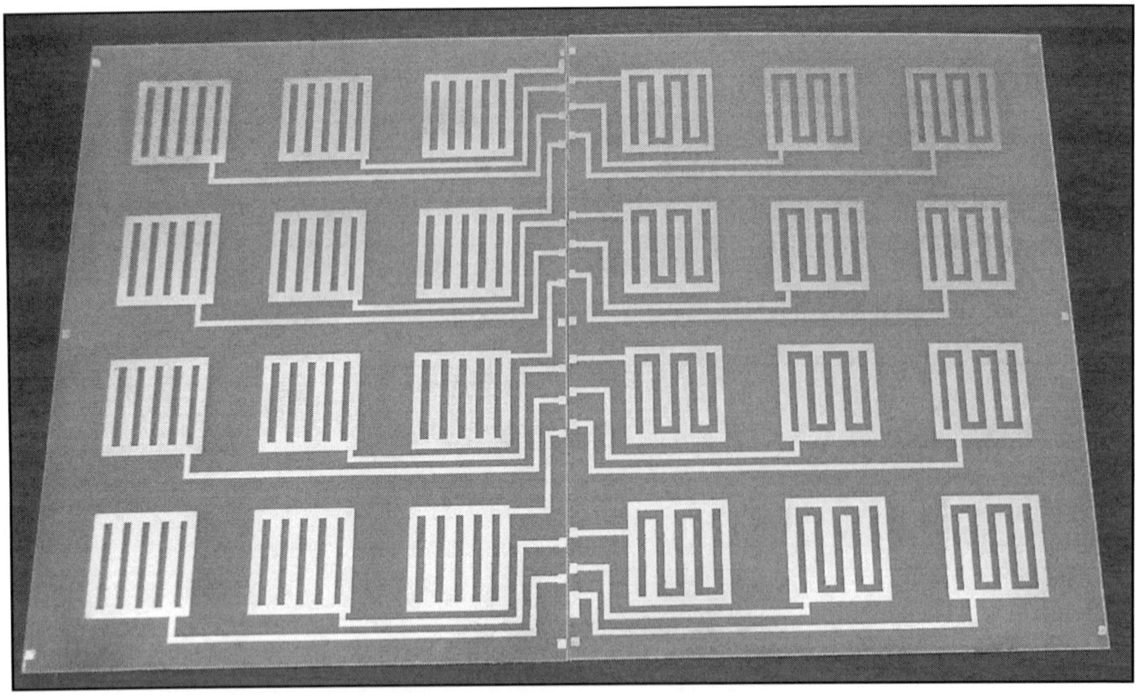

The electrode plates with the etching process complete.
Left, oxygen; right, hydrogen

Remove the resist

When the etching is complete, remove the blue colored resist by scrubbing it off with a Scotch Brite® type pad under running water. As you scrub, the copper trace will reveal itself. Remove all the resist and the plates will be done.

These two plates are the positive and negative electrodes and gas flow fields for the membrane electrode assembly.

Routing the flow fields

The flow fields can be routed either on a drill press or a small hobby milling machine with a 1/8" router bit. If you use a drill press, attach a table to set up a fence, so that you can move the circuit board against the fence as you rout the gas slots.

It's a good idea to purchase extra circuit board to practice on to get used to milling this particular material. It is easy to work with, but getting a feel for it before you work on the actual piece for the fuel cell can be very helpful.

If you don't have a drill press or milling machine, you could ask a friend, or some enterprising high school shop student to do the job; or take a class at a local tech school and bring your circuit board with you as a project. Another option is to take the circuit board to your local machine shop and have them rout the slots for you.

Whatever you do, get extra circuit board for either you or someone else to practice on before the final pieces are cut. Even the most experienced machinist may not have any experience working with this material.

Because of the nature of the thin copper cladding, be sure that the router does not gouge the material. The 1/8" router bit (#60719 from Micro Mark) I have indicated seems to work best, although other kinds will work. Be sure to supply this router bit to anyone who does the job for you unless they are confident that the router bit they will use won't make the copper edges too ragged.

Using a drill press or milling machine

If you use a drill press, I suggest using an XY table. With an XY table you can just clamp the piece to the table and turn a wheel which moves the piece through the cut while you hold the spindle down. If you do not have an XY table, use a fence, and guide the circuit board through the cut with your hands, holding the piece securely. You will need to have someone else lower and raise the spindle for the cuts as you guide and hold down the work piece, so that it does not vibrate or pull up during cutting. When cutting the

A milling machine set up to mill the oxygen electrode plate

circuit board material, move the piece slowly through the cut with a little pressure and let the router do most of the work.

I have done this both ways and I can assure you it is a lot easier with an XY table on a drill press, and easiest with a milling machine. Either type of machine must have enough throat depth to move the 8"x 10" piece around and position it to complete all the cuts. Be sure to check this out – some machines do not have enough throat depth for this project.

Setting up

Most of the work in milling is setting up, that is, planning how to make the cuts. Before you start routing the flow fields, know exactly how you will proceed to the very end, and the milling itself will be an easy task. Place a piece of Plexiglas®

between the circuit board and the table to act as a cutting buffer, and use some small pieces of Plexiglas® on top of the piece to protect the its surface from the clamps, if you use clamps, as in the photo, previous page.

Controlling the depth

Rout all the way through the circuit board. The router bit must be set at a depth that is just a little more than the depth of the circuit board to make a complete cut. If you use a drill press, there are two adjustment nuts on a screw that act as a drill stop or depth stop. Again, set them for the bit to just pass through the circuit board and into the Plexiglas® a little bit. If you use a milling machine, there is also a drill depth stop that must be set before milling proceeds.

When everything is set up, simply pull the lever down and the router will drill into the board. With a milling machine, the router can be locked into position, but with a drill press you must keep pressure on the handle to make sure it stays down, as there is a spring which brings it back up if you do not hold it.

If you are using a drill press without an XY table, this is when you need to have someone hold the lever down while you guide the piece with your hands to route the slots. To rout, simply line up the router bit on the 1/8" spaces between the copper traces of the electrode fingers, set up the fence for a cutting guide and push the piece (on a drill press without XY table) along the slot space.

If you are have an XY table on a drill press or a milling machine, simply turn the wheel to move the clamped piece through each cut. At the end of each slot, lift the router from the piece. Align the piece for the next cut and reposition the fence, if you're using a fence; or if you have an XY table, move the wheel to reposition the piece for the next slot cut.

With smaller milling machines, and drill presses with an XY table, you still have to occasionally reposition the piece and reclamp to make all the cuts. If you use something larger than a hobby milling machine or small drill press, you may not have to reposition your work at all.

When working with the hydrogen side plate, you can move from one cut to another within each electrode because they are connected by 90° angles. You do not have to lift the router out, just change direction. For a drill press without an XY table you will need two fences; or if you only have one fence set up, stop the machine and reset the fence to move the work piece in another direction. With a fence on a drill press, it may be more convenient to set up the fence on one side of the press, drill all the vertical slots on the piece, then change the fence and drill all the horizontal slots on the hydrogen side. Remember to rout the slots for the screws and the edge connectors.

Experiment on your test pieces to see what works best for you.

The plates after milling. Oxygen, left; hydrogen, right

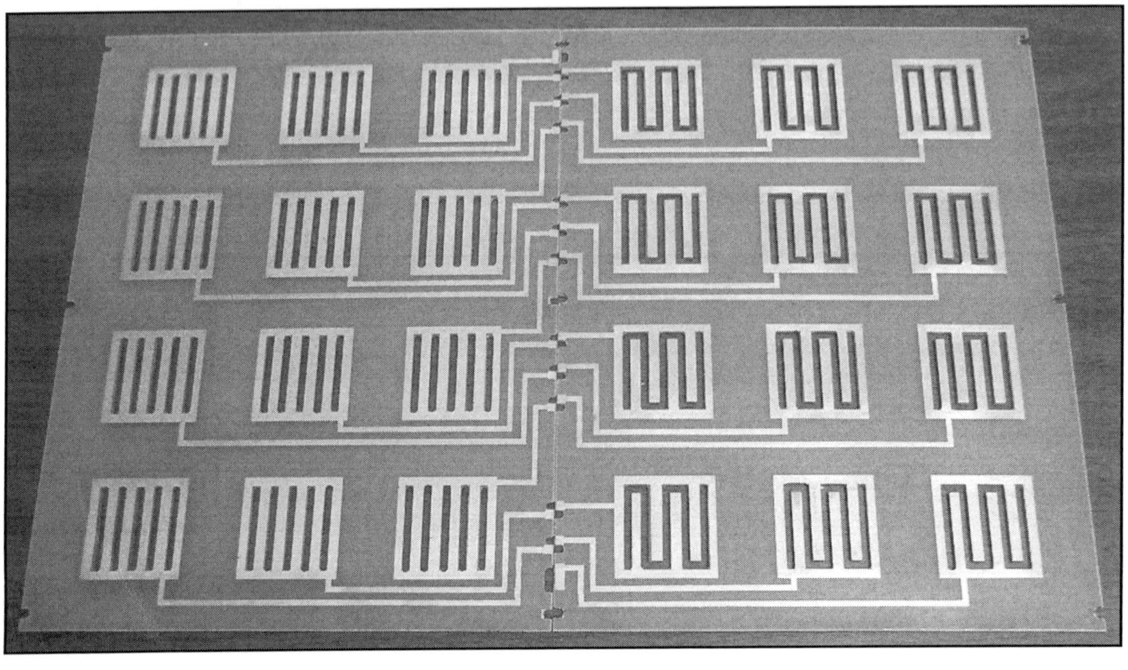

Smooth the edges and clean the plates

After routing the gas flow fields, smooth all the copper edges of the electrode fingers with a very fine file, and/or cut a piece of 1500 grit sandpaper and smooth the edges. It is important to have smooth edges. Parts of the copper trace will be laid against membranes, and rough edges on the flow fields can scratch or puncture a membrane. After smoothing the edges, clean the copper traces again using a fine Scotch Brite® type pad and then rinse with water.

Materials for plating the circuit

Copper corrodes quite readily and under fuel cell conditions can deteriorate rapidly. To ensure that the surface remains conductive, a plating coat of a less reactive metal is recommended. There are several options for coating the copper surface to avoid losing the capacity of the cell. One is to plate the copper surface with Tinnit™. This is a tin base that will protect the copper coating.

Another option is to plate the copper surface with nickel; or, plate the copper surface with nickel and then over plate it with a thin layer of gold.

The goal is to have as little resistance in the circuit as possible so that you can get as much current from the cell as possible. Copper is an excellent conductor of electricity with a low resistance of about 1.7 micro ohms-centimeter at 20°C. Unfortunately it oxidizes and tarnishes, and this deterioration will impede the power output of a fuel cell after a while. Gold does not oxidize readily and is quite conductive, with a resistance of about 2.4 micro ohms-centimeter at 20°C. Even though it does not have as high conductivity as copper, gold will survive fuel cell conditions for a longer period of time. Nickel is highly resistant to corrosion and is an excellent metal to use, but it has a higher resistance (about 7.8 micro ohms-centimeter). Another option is tin plating which has a resistance of about 11.5 micro ohms-centimeter.

So, tin plate, nickel plate, or nickel plate with additional gold plate are acceptable options, but the most desirable is the nickel-gold plate. The straight nickel plate would be the second best and the tin plate would be the third choice.

For this project we decided to put a nickel strike on the copper, and over plate it with gold to enhance conductivity. This took more work and was a little more expensive, but it enhanced the performance significantly so it was worth the effort.

Brush plating

The easiest method to plate these circuits is brush plating. Brush plating is simple electroplating with a brush plating wand. The wand has an absorptive tip which is dipped in the plating solution, and then rubbed and stroked onto the surface to be plated. The pen or wand is connected to the positive pole of a battery or power supply, and the work piece to be plated is connected to the negative pole of the battery or power supply.

As the pen is gently rubbed on the object to be plated, a thin film of metal is deposited on the surface. If you plate the electrode surfaces with gold, plate them with nickel first. If you simply plate the copper with gold, the gold is absorbed or migrates into the copper over a short period of time. To prevent this, a nickel plate base is necessary.

Plating kits

If you have never brush plated before, you may wish to purchase a inexpensive brush plating kit from Caswell plating. They have Plug

N Plate® kits which come with a plug in power supply (a simple 300 milliamp wall transformer), a brush plating wand, plating solution, and instructions.

They have a nickel Plug N Plate® kit which includes plenty of nickel solution to cover the plates. If you choose to use gold also, purchase a small bottle or two of gold plating solution. The nickel plating kit wand and power supply will work with the gold solution. An added benefit is that if you order the kit, you can also get technical support for using it.

Applying nickel plating over the copper layer

To apply tin plating, you can use a Plug N Play® kit, or use an electroless immersion plating solution such as Tinnit™. You could also have a plating shop do the plating.

Applying gold plating over nickel plating

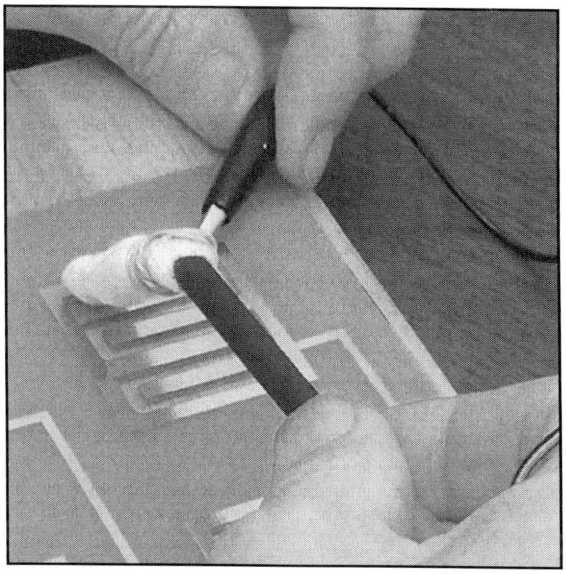

Every part of the metal circuit should be plated except for the tips of the series edge connectors. These will be tinned to be soldered with a soldering iron. If you use Tinnit® for plating, you can plate the tips also. Other metals require different types of solders, so it is best to leave the tips unplated where you will solder the tabs to ensure good connections.

Tinning the series edge connectors

Tin the edge connectors on each of the electrode plates.

The next step is to tin the tips of the series edge connectors on the oxygen and hydrogen electrode/flow field plates. There are twelve contacts to be tinned on each plate, eleven for the series connections and one for the take off on one end of each plate.

To tin, take a 60 watt soldering iron and touch it to the solder. The solder will stick to the surface of the iron tip. Then, rub the solder off the tip onto the connector.

Solder 1/4" tabs to the edge connectors on one of the plates.

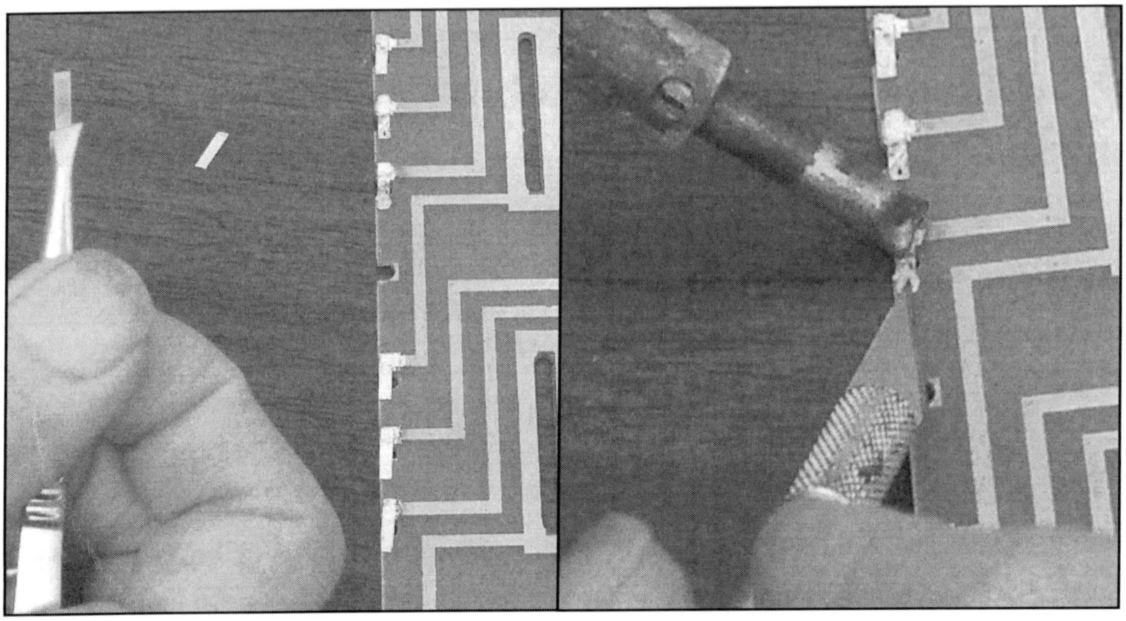

After soldering the tabs to the plate, test for continuity.

Preparing the tab connects

To connect the cells in a series, use tab wire. You can either purchase the tab wire or cut it from copper foil. Use any thickness you like, but it should be flexible enough to work well with this circuit. Generally, a .005 thickness is sufficient. Cut eleven tabs 1/4" long, or, a tad longer as there is a small surface bend to accommodate between the two plates. Don't cut the tabs too long, but don't cut them too short because you want as much contact surface as possible. Tin the tabs on both sides. Solder the eleven series connect tabs to one of the plates.

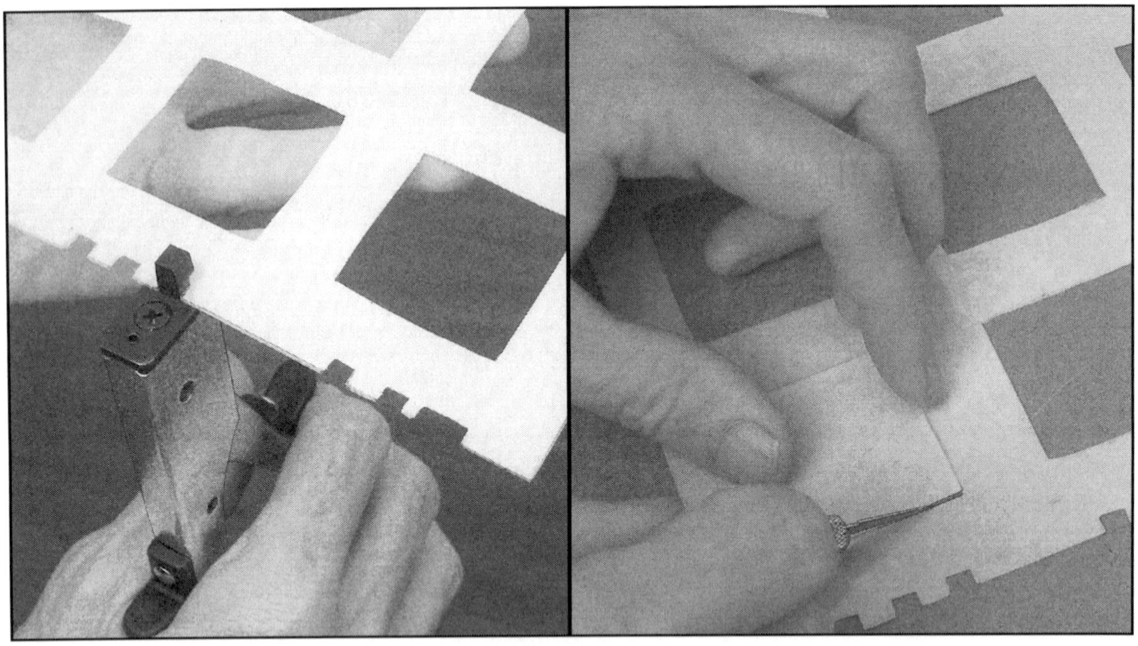

Preparing the silicone hydrogen gasket for the MEA

Making the electrode gasket

From either .010 or .020 silicone rubber, cut the hydrogen gasket with an Exacto® razor knife using the template (see page 205) as a guide. This thin silicone material comes sandwiched between two pieces of Mylar®. The Mylar® will be used to mount the MEAs (membrane electrode assemblies). When you remove the silicone from the Mylar®, the silicone will shrink, so do not cut it while it is sandwiched between the Mylar®. Remove it from the Mylar® sheets and let it sit for a few minutes to shrink to its working size.

The silicone hydrogen gasket

Making the surrounds

Cut the two Mylar® sheets using Template 3 as a guide. It is very easy to over cut when using the Exacto knife, so go slowly, and use fresh blades to be sure to produce clean cuts.

Inserting the membranes

Exercise care when handling the MEAs. It's a good idea to wear cotton gloves so that you do not get any oil, sweat or other contaminants on the membrane and catalyst. On some purchased MEAs, the anode side is marked as you can see in the photo. All anodes must face the hydrogen electrode plate.

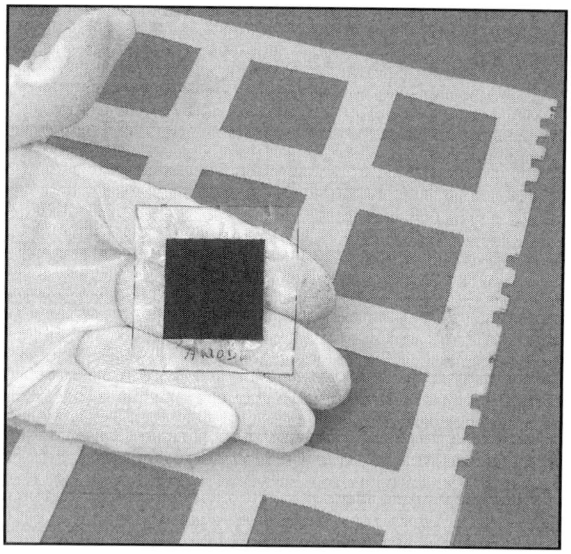

MEA (membrane electrode assembly), Mylar® sheet in background

Each MEA needs to be glued and sealed onto the surface of the Mylar® surrounds. A good seal must be formed when attaching each MEA so that gases from one side cannot travel to the other side. A bad seal will result in a non-functioning fuel cell.

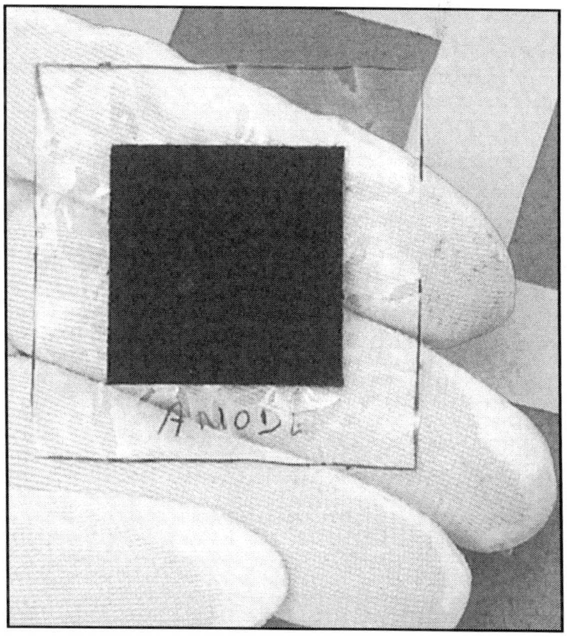

Before gluing the MEAs, center each of the MEAs over the holes they will be mounted in

Trim the ionomer so that the individual MEAs do not touch any of the other MEAs.

and observe whether the ionomer of any membrane touches the ones next to it. Be sure that the ionomers do not touch. If they do, trim them down so that they do not come into contact with each other. Trimming will negate the possibility of ionic cross conduction in the membrane, which can reduce the energy output of the fuel cell stack.

Although the individual membranes must not touch each other, do not trim off too much of the ionomer because there should be as much contact surface for gluing as possible. The larger the contact gluing surface, the more likely the membrane will be properly sealed.

Once the spacing has been fine tuned, the membranes can be glued. Lay one of the Mylar® sheets in front of you and apply silicone adhesive to the glue area on the Mylar® for the first membrane to be glued. Pick up the membrane and very carefully center and place it, being sure not to get any adhesive on the gas diffusion layer (the dark colored active surface area). Press and smooth the ionomer to the Mylar® surface so that it is well secured and soundly attached. Make sure to use enough silicone to cover the glue area very well, but not so much that it will ooze out beyond the glue zone. This needs to be a clean job. If any silicone bumps are left and they dry, they will interfere with getting a good gas seal. Make everything as clean and smooth as possible.

When you press the ionomer to the Mylar®, some of the silicone will ooze out on the edges. Smooth it out immediately, because once it begins to harden, it will be extremely difficult to remove it without damaging the membrane.

After the surface facing you is clean and smooth, turn the Mylar® sheet around and smooth out any oozing surface on the other side. When you turn it around, never place it on the same piece of paper as you might accidentally get some of the silicone on the active surface of the MEA. Use a fresh sheet of paper each time you turn the piece. You will go through a lot of paper but the MEAs will come out unscathed. If any part of the active area of the MEA is covered, it will reduce the output of that cell, and in a series connected stack like this it will also reduce the output of the other cells. After completing the insertion of the first MEA, complete the other eleven.

The completed MEA sandwich: 12 MEAs mounted between two sheets of Mylar®

To complete the MEA sandwich, cover the surface of the other sheet of Mylar® with a smooth and complete coat of silicone to adhere it to the first piece of Mylar®. Very carefully lay the second sheet over the Mylar® membrane assembly in front of you. Press and smooth the two pieces together to ensure a gas tight seal.

Put a smooth layer of silicone around each MEA where the Mylar® edge touches the membrane to further seal the edges against gas leakage. Do this on all four sides of each of the 12 membranes. Set this assembly aside to dry, placing it so that it will not stick to any surface.

Gas supply lines

Use Template 4 (page 206) to cut the rubber gas supply lines from the 3/32" thick silicone rubber sheet. Cut the rubber to the correct size and then tape the template to the rubber. Use an Exacto® knife or other very sharp cutter to make the gas lines. I used a small Exacto® razor knife and found that I had to change blades very frequently – I ran through a whole pack cutting out the gas lines. Cut slowly, and when the template paper starts to bunch up as you are cutting, stop immediately and change the blade. If the template paper rips you will lose the outline which will make it difficult to cut accurately. The process is easy, but as with other operations in this project, it can be tedious.

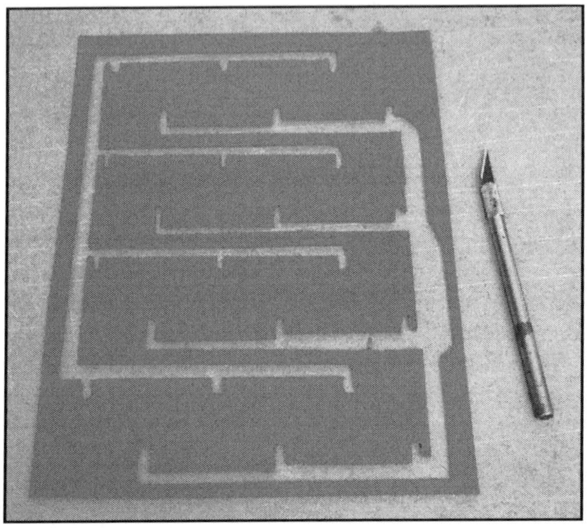

Cut the gas supply lines out of the rubber with an Exacto® knife.

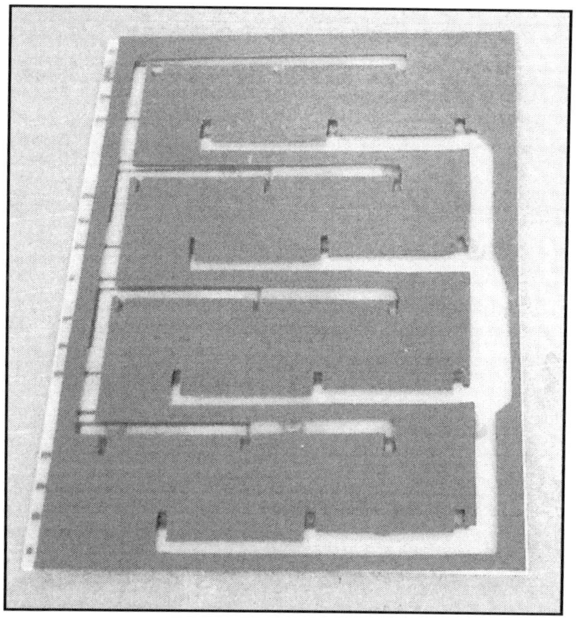

The rubber gasket laid and glued on the back of the hydrogen electrode plate.

Attach the gas supply gasket

Lay the hydrogen electrode plate on your work surface with the metal electrode side down. Lay the gas supply gasket next to it with the side of the gasket that will be glued to the hydrogen plate facing up.

Coat the gas supply gasket with a thin layer of silicone caulking. Pick up the gasket and position it quickly onto the hydrogen plate before the silicone gets too tacky and sets. All

Hydrogen electrode plate with rubber gas feed gasket glued to its back.

of the gas supply lines must be aligned so that they feed into and out of each cell properly. Check and adjust these quickly before the silicone caulking sets. Press and rub the rubber gasket to make sure the gasket is sealed against the electrode plate surface. Set it all aside and let it dry for twenty four hours.

Note the illustration on the previous page, which shows the hydrogen electrode with the gas feed gasket glued to its back, as seen from the electrode side. The gasket is visible through the routed flow fields in the circuit board, and also somewhat visible through the unplated parts of the board. Also note the gas inlet and outlet holes at the beginning and end of each flow field.

Preparing the end plate

Draw an 8"x 10" outline centered in the PVC plate. Align a copy of Template 5 (page 207) for the Pressure Plate inside this 8"x 10" outline. Take a sharp object and press the point into the cross hairs of the 1/4" and 1/2" outlet, and the six holes for the pressure screws to make indentations to mark the PVC plate.

Remove the template, and then drill pilot holes for the gas inlet and outlet with a small size drill; then drill the final holes with the correct size drills. The pilot holes are not necessary, but they can help your drilling to be more accurate. Also, do not drill the final hole too big. There should be as tight a fit as manageable for the barb hosed connector.

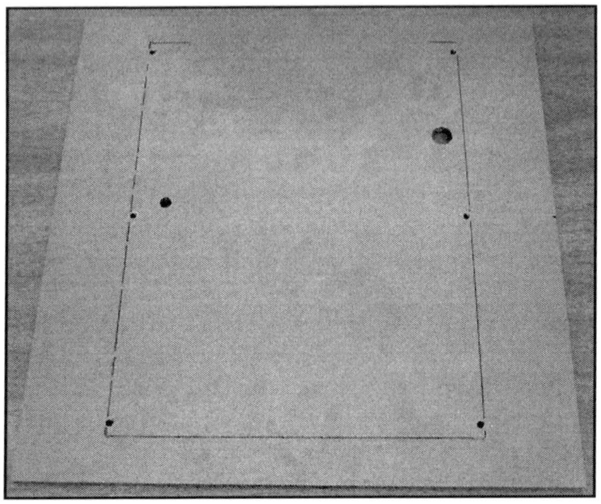

The PVC end plate with holes drilled for the gas inlet and outlet and the pressure screws

Drill the six $^1/_8$" holes for the pressure screws.

Aligning the plates

You can use the PVC end plate as a base to align the electrode layers. To do this, three of the pressure screws are inserted through their holes in the end plate, but in the opposite direction of the way they will be installed when the fuel cell is completed.

Put the hydrogen electrode plate (with the gasket glued to it) on the PVC plate with the metal electrode surface up, and the gasket side down; and align it with the guide screws. Wash the .010 or .020 silicone gasket material with water and dry it, then lay it on the hydrogen plate. Align it with the electrode cells (see illustration, top right).

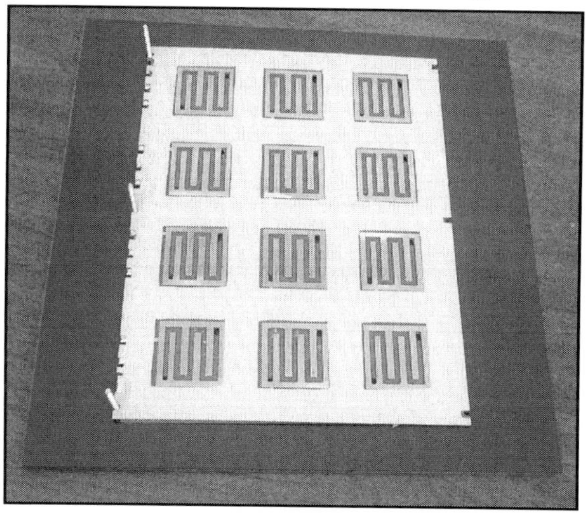

Three pressure screws are inserted through the PVC end plate to align the electrode layers.

Place and align the Mylar® MEA on top of the silicone electrode gasket (bottom photo at right).

Then, as shown at right, take the oxygen side plate and place it on top of the Mylar® MEA, with the plated electrode surface face down, and align. Add the other three screws and tighten them down slightly with butterfly nuts, as shown in the photo.

Ready to solder with all the electrode layers in place.

They should be tight enough to hold the layers in place, but should not exert downward pressure on the layers, as this could interfere with soldering the connectors.

Soldering the series connections and power takeoff tabs

Since the series interconnect tabs are already soldered to the oxygen plate, all you have to do is solder the tabs to the hydrogen plate to complete the series connection. To do this, put the soldering iron tip in the space where the tab is and push down until the solder melts and a bond is formed. A sixty watt iron with a chisel tip usually works well and is the correct size to fit in the space allotted. This soldering operation is tricky because you do not want to desolder what you already soldered.

Soldering the connectors together

If you don't have much soldering experience and are not confident about this operation, you can make tabs that extend away from the plates and connect them together away from the fuel cell (see illustration below). Instead of using 1/4" tabs as shown on page 174, make the tabs a little longer, and cut 22 of them. Then, solder a tab to each series connection on both plates, instead of just to one plate. After the component layers are aligned as described before, solder the tabs together away from the edges of the plates. This may be an easier method; however it will not be as neat and compact looking.

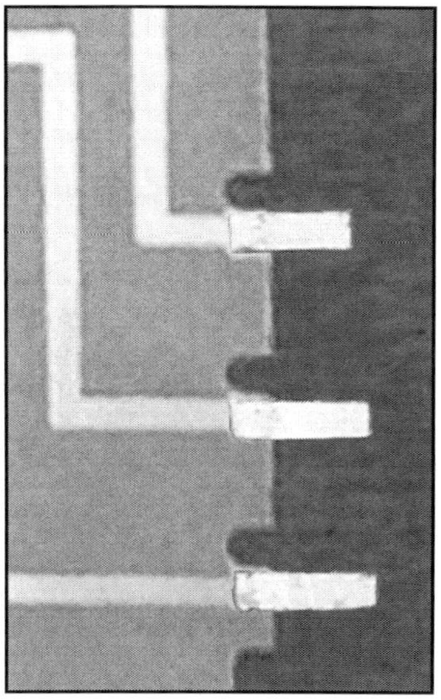

Alternative way to make the connections

Whatever method you use, be sure that each connection is good. If one of them is not, the fuel cell won't work.

When you have finished soldering the series connections, the positive (oxygen) side of each cell will be connected to the negative (hydrogen) side of the next cell in series, so the voltage from each cell is added to get the total output of this planar fuel cell stack.

Next, cut two pieces of 1/4" wide by .005 thick tab ribbon. These tabs are the power take off tabs. They can be cut long, and trimmed to the desired length when the fuel cell is finally assembled. One tab is connected to the hydrogen side take off connector and the other to the oxygen side edge take off connector. You will be able to solder one of the take-off tabs with the compression screws on, but the other take-off is on the other side and you will need to solder that when you take the sandwich off the PVC plate and turn it around. Tin the tabs before soldering. Cover a portion of the power take-off tabs with a liquid electrical tape to insulate them from touching the compression screws and aluminum channel bars.

Carefully loosen the compression screws and gently remove the sandwiched layers from the PVC plate, solder the other take-off tab, and set them aside. The sandwich may tend to fan out on the edge opposite the edge with the soldered connections. This is not a problem and is to be expected. When the sandwich is finally laid in between the PVC plate and the channel holds, and the edges are sealed, the layers will be evenly compressed.

Preparing the pressure braces

Pressure braces ensure even pressure across the electrodes so that they connect well with the MEAs.

Aluminum is readily available and inexpensive, but aluminum in bar or rod shape has too much bend for this application. We chose aluminum channel for the pressure braces. Because of its structural shape, the channel stock has stiffness that flat pieces of thicker metal bar do not have.

Cut three 12" pieces of 5/8"x 5/8" aluminum U-channel for pressure braces. Drill 1/8" holes on either side of the channel to match the holes in the PVC plate so that they mate correctly.

Aluminum U channel stock, cut, drilled and marked for the pressure braces

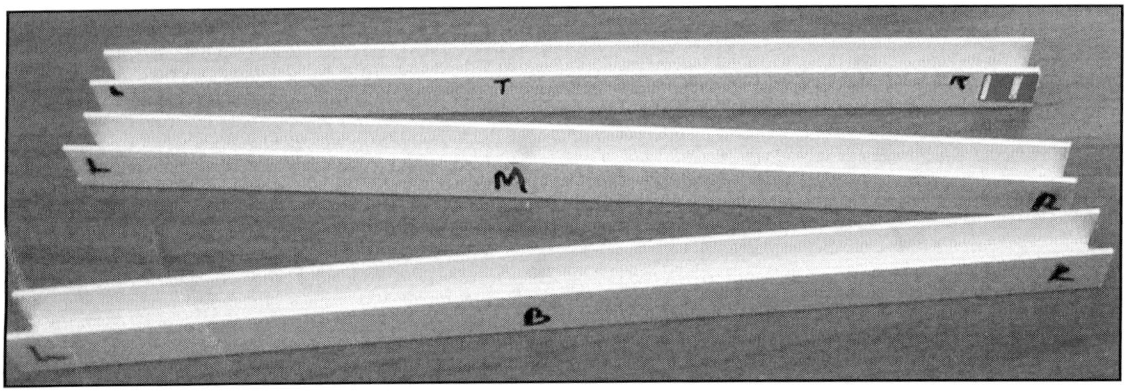

Debur all the edges and then wash the channel to remove any aluminum powder from the cutting, drilling, and deburring process. Any metal powder that later finds its way onto the membrane surface via the air ports could be detrimental to the operation of the cell.

Installing the gas port connectors

Prepare the barb connectors for installation in the gas port holes in the PVC plate. Cut the connectors so that they will be flush with the inside of the PVC plate. Apply epoxy glue to the inside rim of the drilled holes in the plate, then apply epoxy to the outside rim of the barb connectors that will contact the PVC plate. Insert the barb connectors into the holes and then apply more epoxy to the top and bottom edges where the barbs contact with the PVC plate to assure a good seal. Clean off any excess epoxy, but also leave some to provide ample seals. Avoid getting any epoxy inside the barb connector. This would block the gas flow through the connector. Lay the PVC plate aside and let it dry.

Barb connectors for gas ports

Final assembly

Place the electrode assembly onto the PVC plate and align. Take a piece of channel and position it, insert the screws in the holes with the screw heads inside the channel. On the opposite side of the plate, place a washer around the screw, put a wing nut or other nut onto the screw and twist it down to hold the channel and the plate together. Finish by applying the other two pieces of channel to the plate.

Completed fuel cell, with the back side of the air plate visible

Tighten the wing nuts by hand, and do not let the plate bend from over tightening. The plate should be straight across without the slightest induced curves.

Keep the take off tabs clear of the screws as you insert the screws. In this design, one of the take off tabs comes very close to one of the screws, so it is important to move it around the screw if necessary and be sure that the tab is coated with an insulated material such as silicone, liquid electrical tape or rubber dip compound as mentioned earlier to prevent contact between the ribbon and the screw or channel. For more rigidity you may want to insert stiff metal strips under the channel at

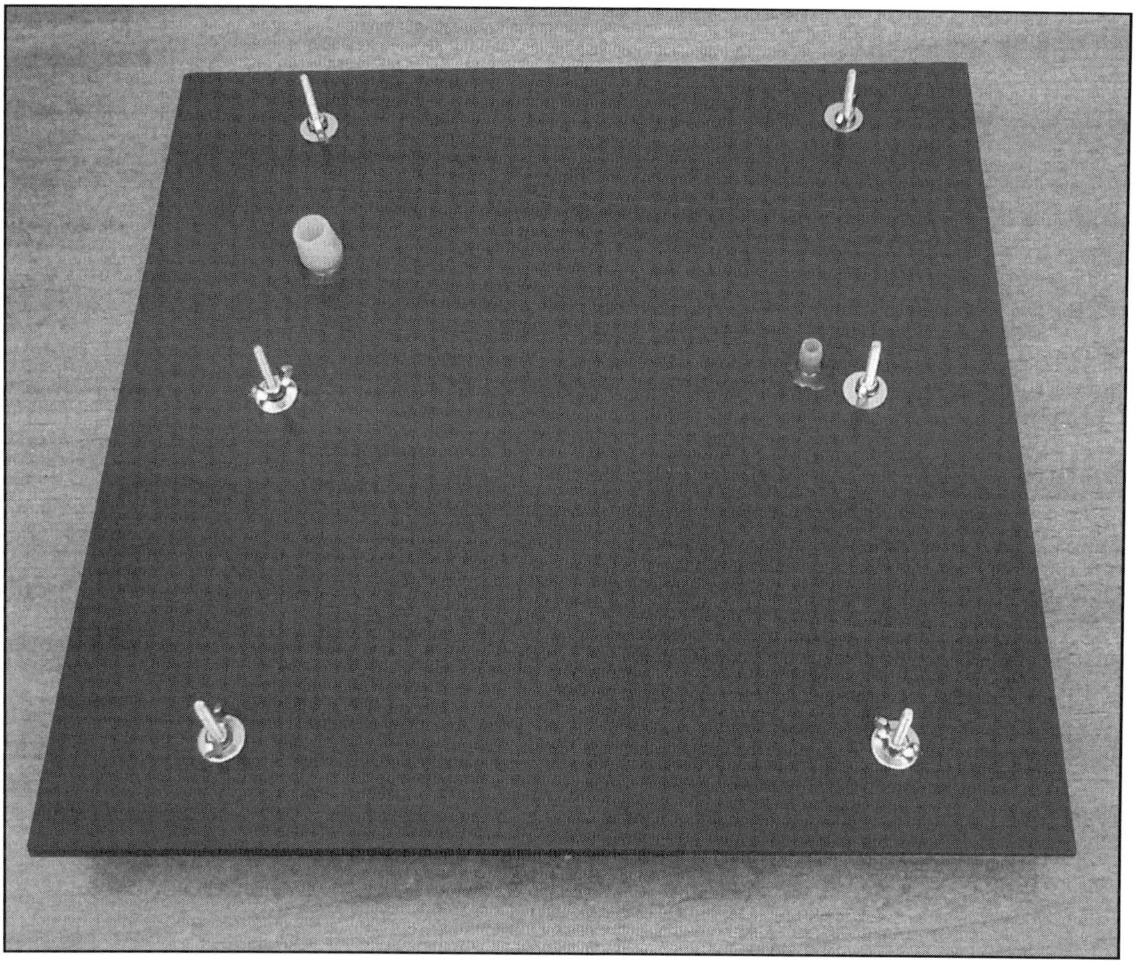

Completed fuel cell showing the hydrogen inlet and outlet

several points to insure more even compression. If you do this, make sure you do not cover the air slots on the oxygen side with the metal strips.

Apply a coat of silicone to the edges of the electrode assembly. Completely cover the edges and where they meet the end plate to ensure a gas tight seal. Then, smooth the edges with your finger or a razor blade to make a nice clean, even looking finish. Set the fuel cell aside to let the silicone dry.

Prepare the stack for testing

Inspect the fuel cell stack to see that everything is in order and the pressure screws are tightened. If you have doubts about the silicone coverage on the edges, apply more to ensure a tight seal, then let it dry for 24 hours before testing.

You can attach a small piece of hose to the gas exit port barb connector and connect a valve to control the amount of gas that exits the cell. This exit port gas regulation can help to fine tune and tweak the cell output. You should be able to "dead end," that is, go from a completely closed port and open the valve in increments for more exit volume. Please note that you should not dead end the cell if there is no storage container with pressure relief as outlined in this book, or some such other exit for the gas. The GPS system has a check valve in the design which prevents the gas from flowing backwards in the system. This can cause excessive pressure build up in the fuel cell stack which can cause gas leakage, or rupture the membranes which will destroy the stack.

Purging the stack

To purge the stack of air before introducing hydrogen you can use bottled gas, or purchase a good size helium balloon. When the balloon is filled, have it left tied loosely so that you can open it easily. Put the nozzle of the balloon on the exit port of the fuel cell and let the gas expel into the fuel cell and out the entry port. At the same time, have the regulator turned on to provide hydrogen and insert the hydrogen hose onto the intake of the fuel cell. Remove the balloon and the stack is purged.

We have not used inert gas purging, since our system is experimental and set up for short runs. For a permanent system, if there is a need to purge the stacks, inert gas purging and a built in purging system should be considered.

Testing the fuel cell stack

To test the fuel cell, your hydrogen production system needs to be set up and running, or you need another source of hydrogen available.

First, connect the gas supply line from the regulator to the fuel cell entry port barb connector. As a general practice, always have some sort of flashback arrestor between the regulator and the fuel cell stack. This can be a small flashback bubbler that you can construct and install, or you can use a commercial flash arrestor.

Use a multimeter or voltmeter to do an open circuit voltage test. Connect the leads from the meter to the positive and negative take-off terminal tabs on the fuel cell.

If you are using the GPS system design in this book, open the regulator that is connected to the electrolyzer/storage system to allow hydrogen gas to flow to the fuel cell. If there is a gas exit valve on the fuel cell, be sure to open it for full gas exit. Then, you can diminish the gas flow in increments if you wish during testing.

Allow the gas to flow to the fuel cell for a minute or two, and then take the reading from your voltmeter. Readings should be from 15V through 9.6V. After a while the voltage will drop and plateau at various points. You can expect the cells to settle out and maintain a constant voltage of around 7.5 to 10.8 volts. This will vary, however and you will have to test your particular fuel cell stack to determine its characteristics.

This small power stack was designed to produce 6V at between 1 to 2 amps under load. The current output depends much on the quality of the membranes used.

To test the short circuit current, simply change the multimeter setting to the current reading or use an ammeter and connect the probes to the same terminals that were used to get the voltage reading.

Trouble shooting

If you do not get any readings, you have just entered fuel cell builder's hell. This is a place where your shoddy previous work (or somebody else's) comes back to haunt you, and you are forced to contemplate and work to fix your mistakes.

So, where could the problem be and what is the best way to troubleshoot your system? The first rule of troubleshooting is to prevent trouble in the first place as much possible. Do not rush while constructing the stack. Test, check and recheck as you go along, and do things right the first time so that you don't have to troubleshoot. Then, if there are still problems, you can be confident enough to rule out many possible causes of trouble, and zero in on the real cause and fix it.

Make a troubleshooting check list to have in hand before testing the fuel cell stack, and follow the check list regimen always, because if you don't, it is very easy to get confused.

In a system of this nature there can be many different problems. First, check for the obvious. Are the multimeter probes connected? Is the multimeter working properly?

Then, look at the system as a whole. A simple visual inspection of a system can sometimes show the problem immediately. Is everything connected properly? Are there any gas leaks causing pressure failure or loss of gas? Are valves connected properly, and turned on or off, at whatever setting they should be for the system to work?

If you are sure the system is getting gas, then consider closing the valve in increments on the outlet of the fuel cell to see if that produces any response on the multimeter. Do this until it is dead ended. After a short period of time, if there is no meter response, open up the valve again and move on to another test.

Next, check the fuel cell stack itself, and whether or not there is gas leaking between the plates, etc. If there is a leak here, it must be resealed.

Since this stack is connected electrically in series, if one of the PEMbranes is not working, it will shut down the whole stack. This is why each membrane should be tested prior to installation in the stack. However, there can also be damage to a PEMbrane during construction of the fuel cell. Another possibility is that the connections are not soldered well and you will have to inspect the tabs to see if that is the case. More than likely there will be no problems.

To test individual PEMbranes prior to assembling them in the final MEA, you can make a small test fixture. Construct a single slice fuel cell, such as the easily constructed L78 detailed in *Build Your Own Fuel Cells*. To use the L78 as a test cell, don't glue the Mylar® sheets together, just insert the PEMbrane between the two Mylar® sheets along with the gaskets and other layers of the fuel cell, and hold the layers together for the test with insulated binder clips, or use screws as outlined in the text. You can of course, improvise on this as you wish.

Designing fuel cell stacks

The L79 is a planar variation of some of the construction ideas involved in the L78 stack, as presented in *Build Your Own Fuel Cells*, which also has printed circuit board for the electrode plates. While working with this material, we realized that flow fields could be routed in the circuit board, so that the material could function for a dual purpose: as both a collector plate, and a gas flow field.

As you construct the L79, no doubt many new ideas and configurations will come to mind. After completing this fuel cell stack, you should feel confident enough to make modifications to improve the performance; and you'll probably discover better or more inexpensive materials and methods for fuel cell construction. Once you understand the basics of fuel cell construction, it is easy to come up with many variations.

Fuel cell power supplies

You may wish to build a fuel cell stack power supply with greater voltage and current output.

A good start is to construct a 12 volt 10 ampere power supply, which can be constructed with either planar or block stacks, as discussed in *Build Your Own Fuel Cells*. Whatever design you choose, this project will require 120 MEAs that will deliver around .5 volt and 2 amperes each under load.

Shop around to get the best price for MEAs. There are suppliers listed in the *Resources* section of this book. You can also make your own MEAs, and if this interests you, *Build Your Own Fuel Cells* has a detailed section about this subject.

To construct the power supply based on the L79 planar stack will require ten stacks of twelve MEAs each, and will have an output of 6 volts at 2 amperes per stack. Five series connected pairs of stacks produce 12 volts at 2 amperes per pair. The five pairs are then connected in parallel for a total output of 12 volts at 10 amperes for the power supply (see photo next page, top).

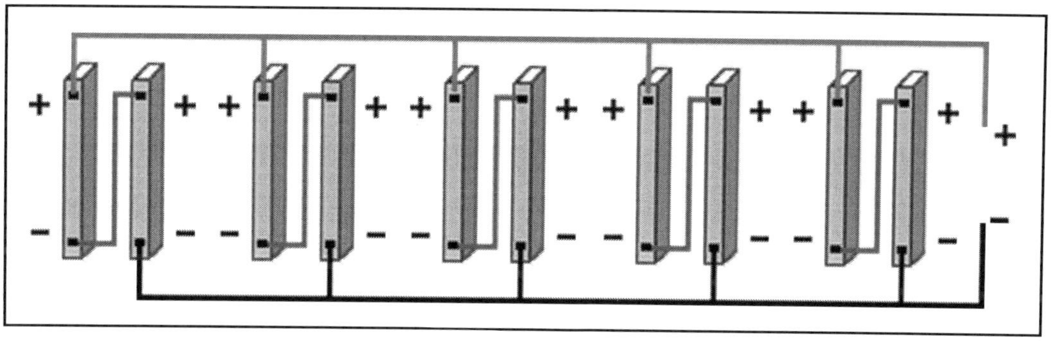

Wiring for 5 pairs of series connected stacks, wired together in parallel.

These planar stacks can be racked in a variety of ways. The illustration below shows a simple tubing manifold system for feed lines and gas exit lines. On the next page is another view of the rack of stacks.

When designing a power supply, consider the number of photovoltaic panels needed, as well as how many electrolyzers. Then, size the gas processing system and storage to provide the appropriate quantity and quality of gas for your fuel cells.

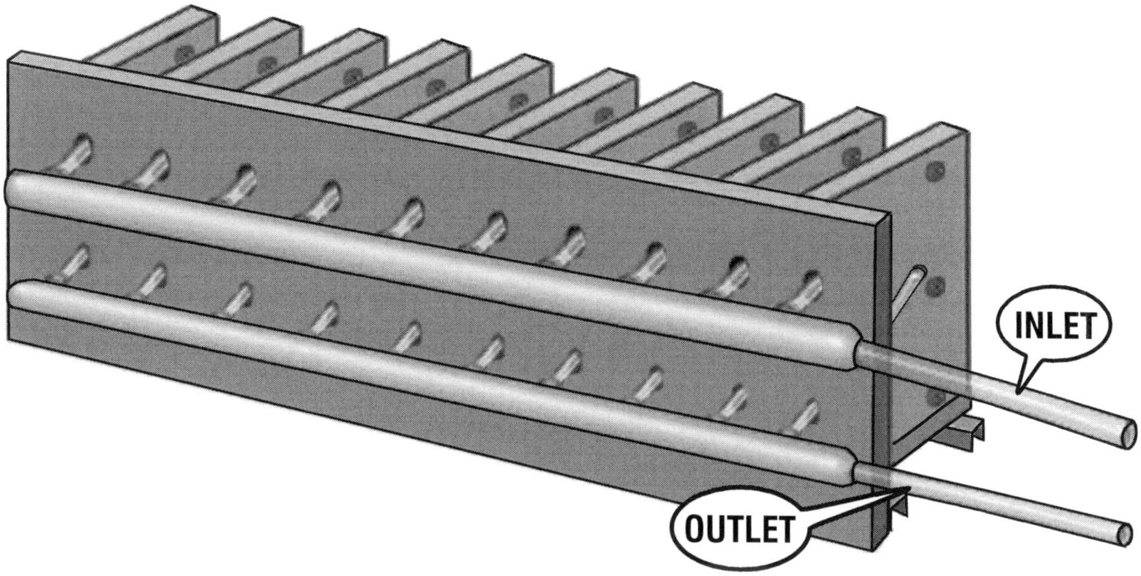

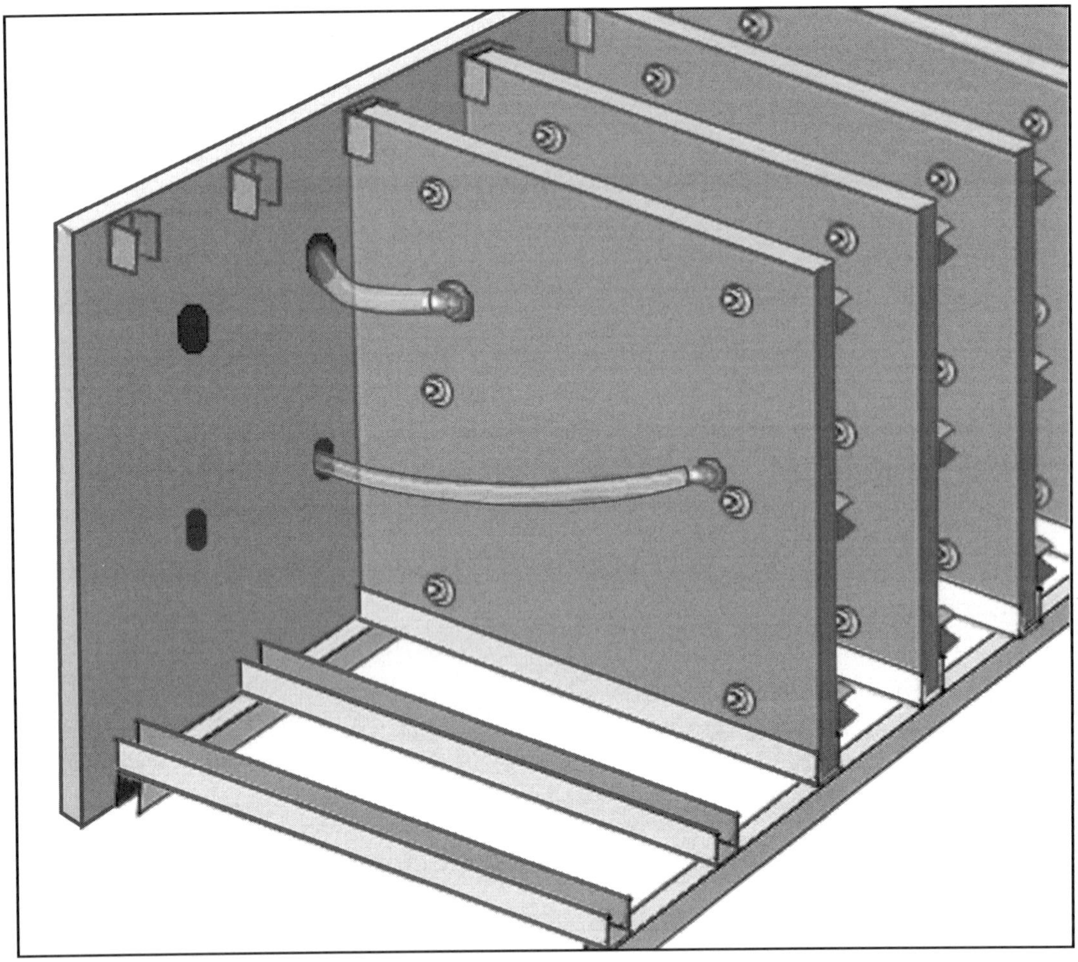

Above and lower left, rack of planar fuel cell stacks

Output configurations

You can take power straight from the fuel cells; or use supercapicitors to condition the output; or feed small rechargeable batteries if you have special applications that require them.

For instance, if there are high amp draws from motor start ups, etc., put a supercapacitor in parallel connection with a 12 volt rechargeable battery, and use this to supply those intermittent load needs adequately. To use a rechargeable battery alone, as mentioned, simply connect the output from the fuel cells to the battery and draw power from the battery. To use a supercapacitor and rechargeable battery, connect the battery and supercapacitor in parallel, connect the fuel cell output to these, and draw your power from the supercapacitor and battery connected leads. With these system additions you will need a diode so that reverse flow does not occur to the fuel cell stack, and fuse the circuit on both sides in case of shorts. A switch, either remote or direct, should be used to connect the power supply with any lines or equipment being powered. If you have AC power requirements you will need an inverter to convert DC to AC electricity.

Running the stacks

In operation, let the fuel cells run for a few minutes before connecting to a load. It would be a good idea to install a voltmeter and ammeter from the fuel cell to output line, and also at other junctions if so desired, so that you can monitor the system at various points. These systems can be totally automated, manually controlled, or both.

Cold weather can wreak havoc with PEM fuel cell stacks for a variety of reasons. It is important to set up the power supply so that it is as well insulated as possible. At the same time, proper air flow must be maintained to feed the cells, and adequate ventilation provided to keep any hydrogen gas from accumulating.

Hydrogen gas monitors that are correctly calibrated on a proper schedule should be used in any power supply system to provide alarms for hazardous conditions. Automatic cut-offs can be used in conjunction with an alarm system to shut down the gas flow to the power supply.

System maintenance and reliability are very important, so be sure to consider this in every step of your design. Engineering a practical and safe fuel cell power supply takes a bit of thought. It is best to start out small and build up slowly, so that your mistakes do not turn out to be costly or dangerous mistakes.

Of course, you can purchase a new or surplus stack and then build the rest of the system around that. If you do buy a unit, find out how easy it will be to replace the parts and perform other maintenance.

Have fun, and always be safety conscious, so that you can have many more days of experimenting and discovery.

Resources

Safety

Material Safety Data Sheets (MSDSs)

You will need to view each MSDS and implement all recommendations and requirements before using, storing, handling and disposing any chemical listed in this book.

It is your responsibility to keep yourself and others safe when working with these substances. These substances are toxic, and some are both toxic and corrosive. You absolutely must wear safety glasses when working with any of these materials. It is good to wear a face shield when working with potassium hydroxide or sulfuric acid.

According to MSDS recommendations, a respirator is required when working with any chemical. Rubber gloves are a must when handling both toxic and corrosive substances listed in this book. Protective clothing per MSDS should be worn. When working with potassium hydroxide and sulfuric acid great care must be taken to protect yourself from these substances.

Some of these substances are toxic – do not eat while working with them. Use a fume hood per recommendation of MSDS. If you do not intend to work with these substances in safe manner DO NOT WORK WITH THEM!!! The author assumes no responsibility for your disregard for safety practices.

Keep all chemicals away from children and pets, and do not allow children or pets into your work area. Do not allow unauthorized persons to enter your work area. Do not allow any smoking or possible source of ignition into your work area when working with hydrogen and oxygen gases.

MSDSs are available on the internet. Simply search the chemical or gas name and "MSDS".

Authority Having Jurisdiction (AHJ)

AHJ refers to an organization or agency that has jurisdiction to make and enforce guidelines and laws for the safe use of certain substances and processes. In most instances there are multiple relevant organizations or agencies, ie, officials and laws at the federal, state, county, and city or town levels. AHJ is not specific organization.

It is necessary for you to find out who the AHJs are for your area before commencing any work in this book so that you can be in compliance with the laws and regulations in effect.

Safety guideline resources

Available on the internet

NFPA	National Fire Protection Association
OSHA	Occupational Safety and Health Administration

Hydrogen Resources

Available on the internet

AHA	American Hydrogen Association
NHA	National Hydrogen Association

Suppliers of Tools and Materials

Available on the internet

McMaster-Carr – Many, many useful items

HMC Electronics – Rosin, flux pens, solder

E. Jordan Brookes Co., Inc. – Tab and bus ribbon

Plastecs – Solar cells, tab ribbon

Ebay and other online sources - tab and bus ribbon, solar cells

All Electronics – Electronic and electrical supplies

Surplus Sales of Nebraska – Electronic and electrical supplies

Circuit Specialists – Printed circuit board, ferric oxide etchant.

Micro-Mark – Hobby milling machines, routers, end mills, and lots of other items useful for fuel cell building

Caswell Plating – Electroplating and electroless plating supplies.

Techniks Inc – Press-N-Peel PCB transfer film

Fuel Cells Etc – MEAs

Electrochem, Inc - MEAs

Ion Power - MEAs

Fuel Cell Earth - MEAs

Templates

The fuel cell stack templates are printed here at 50% of actual size. You can download a PDF version of the templates that will print at actual size at www.solarh.com.

Template 1 – Negative (Hydrogen) Electrode Plate, from PC board

Template 2 – Positive (Oxygen) Electrode Plate, from PC board

Template 3 – MEA Surrounds and Gasket, 2 mylar and 1 silicone rubber

Template 4 – Rubber Hydrogen Gasket, silicone rubber

Template 5 – PVC Pressure Plate

Instructions for using the PDF templates

If possible, set your printer options to print to the edge of the paper, or print the templates to a paper size such as 11" x 17", if your printer has that capability. Because the electrodes and MEA assembly are 8" wide, some printers may not be able to print either the right or left edge for Templates 1, 2, 3, or 5. This will not matter for printing to the transfer film for the electrode templates – they have been aligned so that all of the image needed for the transfer can be printed by any laser printer on the 8.5"x 11" transfer film sheets. For the other templates, if an edge line is missing, you may wish to take a ruler and mark it on the template at exactly 8" wide. Use the template page in this book as a reference. Also, for printing to the transfer film, adjust the laser printer settings for maximum coverage, or the darkest/highest quality option for printing.

Template I

Negative (Hydrogen) Electrode Plate, from PC board

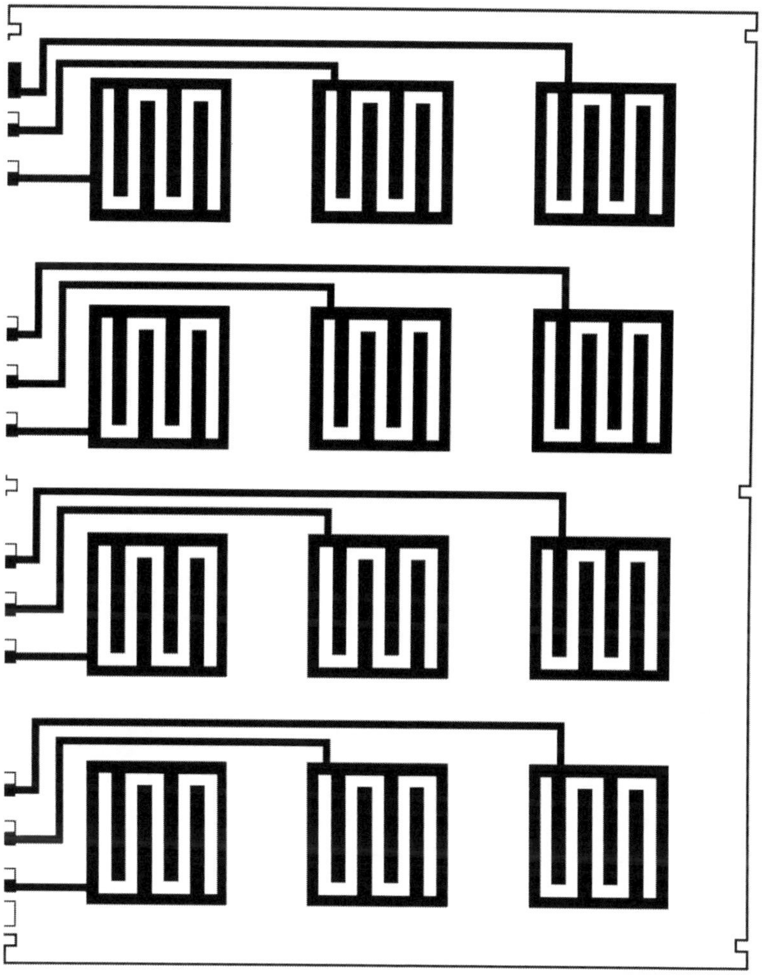

50% of actual size.
Download actual size PDF templates at www.solarh.com

Template 2

Positive (Oxygen) Electrode Plate from PC board

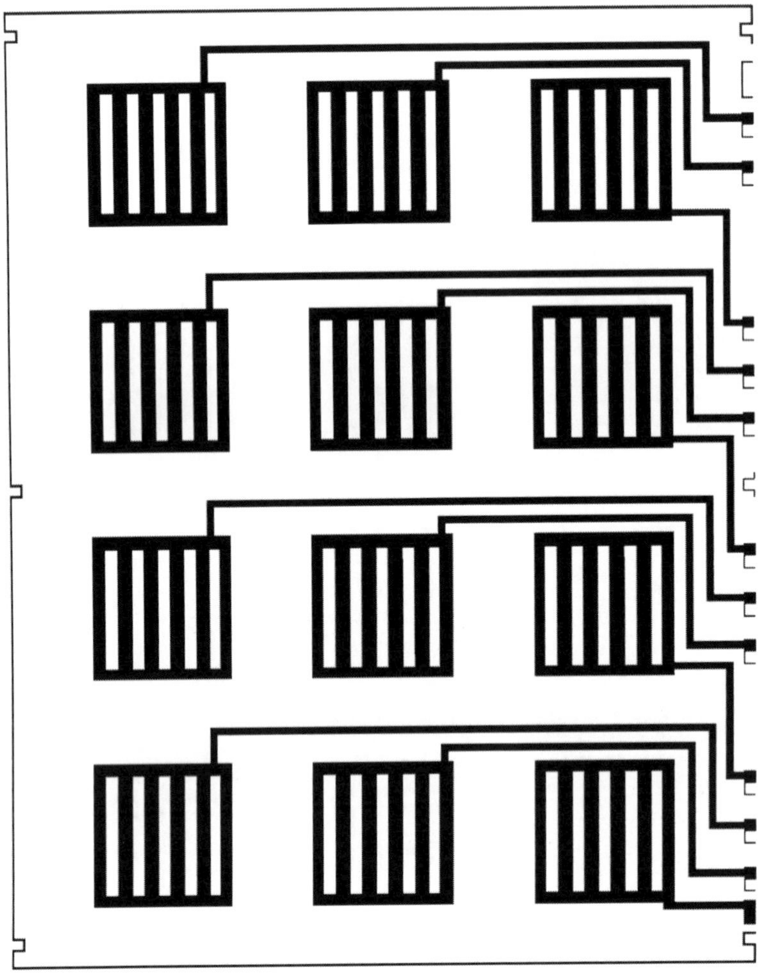

50% of actual size.
Download actual size PDF templates at www.solarh.com

Template 3

MEA Surrounds and Gasket
2 mylar and 1 silicone rubber

Cut 2 from mylar, and 1 from .010 or .020 silicone rubber

50% of actual size.
Download actual size PDF templates
at www.solarh.com

Template 4

Rubber Hydrogen Gasket
silicone rubber

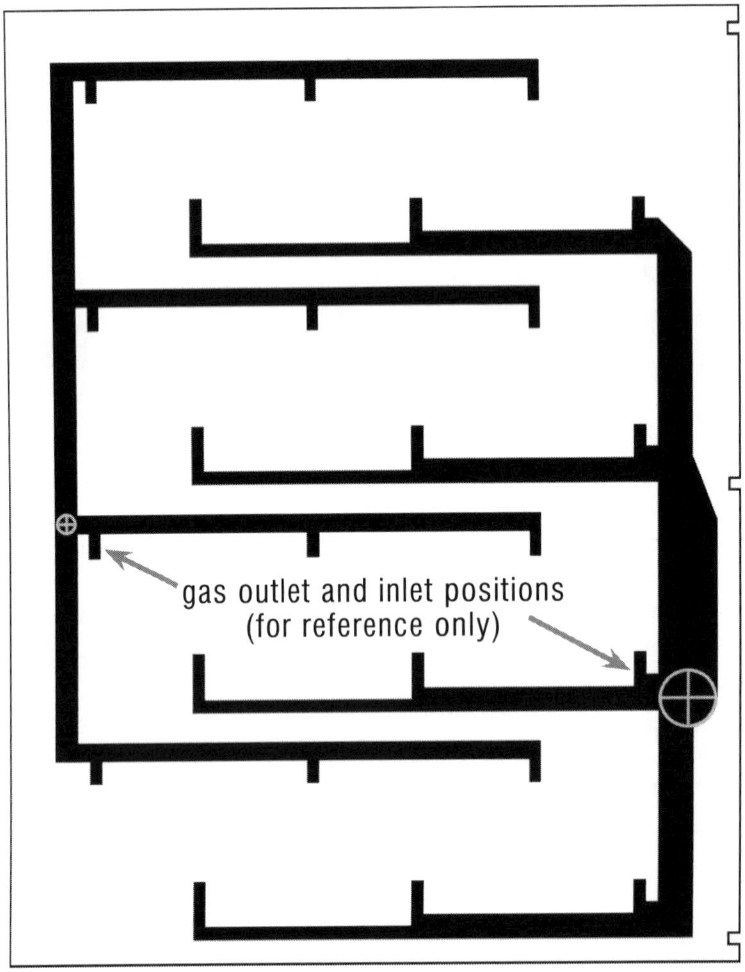

50% of actual size.
Download actual size PDF templates
at www.solarh.com

Template 5

PVC Pressure Plate

Use this template to mark the screw holes and gas inlet and outlet on the side of the PVC plate that will face the rubber gas gasket. Center the outline on the 12" by 12" plate.

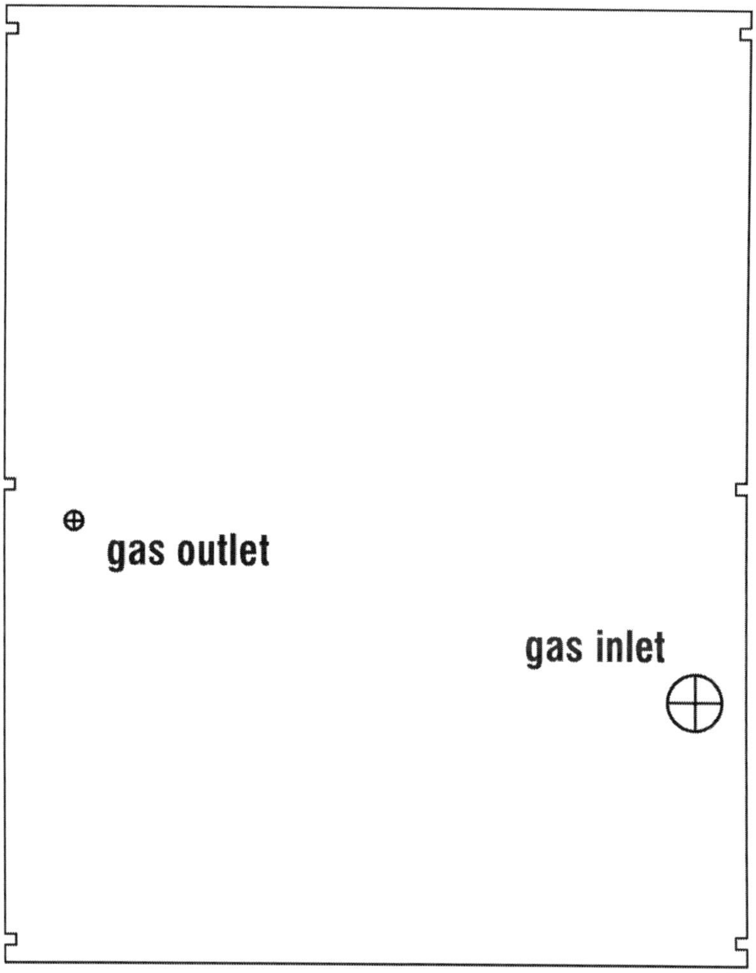

50% of actual size.
Download actual size PDF templates
at www.solarh.com

Build Your Own Solar Panel

by Phillip Hurley

Whether you're trying to get off the grid, or you just like to experiment, *Build Your Own Solar Panel* has all the information you need to build your own photovoltaic panel to generate electricity from the sun. The new revised and expanded edition has easy-to-follow directions, and over 150 detailed photos and illustrations. Materials and tools lists, and links to suppliers of PV cells are included. Everyday tools are all that you need to complete these projects.

Build Your Own Solar Panel will show you how to:

- Design and build PV panels
- Customize panel output
- Make tab and bus ribbon
- Solder cell connections
- Wire a photovoltaic panel
- Purchase solar cells
- Test and rate PV cells
- Repair damaged solar cells
- Work with broken cells
- Encapsulate solar cells

Available in print from Amazon.com and for download in full color PDF ebook format at

www.buildasolarpanel.com

Download a free sample of Build Your Own Solar Panel in full color PDF at www.buildasolarpanel.com.

Solar II:

How to Design, Build and Set Up Photovoltaic Components and Solar Electric Systems

by Phillip Hurley

Now that you've built your solar panels, how do you set up a PV system and plug in? In the e-book Solar II, Phillip Hurley, author of *Build Your Own Solar Panel*, will show you how to:

- Plan and size your solar electric system
- Build racks and charge controllers
- Mount and orient PV panels
- Wire solar panel arrays
- Make a ventilated battery box
- Wire battery arrays for solar panels
- Install an inverter
- Maintain solar batteries for optimum life and performance
- Make your own combiner box, bus bars and DC service box

Solar II includes over 150 photos and illustrations, and a daily power usage calculator. Published April 2007.

Available in print from Amazon.com and for download in full color PDF ebook format at :

www.buildasolarpanel.com

Download a free sample of Solar II in full color PDF at www.buildasolarpanel.com.

The Battery Builder's Guide

by Phillip Hurley

The Battery Builder's Guide is a practical hands-on text that will show you how to make your own rechargeable flooded lead acid batteries. Learn how to recycle parts and materials, how to fabricate battery components and where to purchase the parts, materials and tools you need to build or rebuild batteries. The text covers construction of batteries with Plante (pure lead) and Faure (pasted lead) plates.

Topics include:

- Recycling old lead acid batteries
- Molding battery parts
- Design formulas and tables
- Lead burning
- Techniques and tools for battery building
- Building plate burning racks
- Pasting and forming plates
- Types of batteries such as SLA and deep cycle, and their characteristics and uses
- And more... all illustrated with extensive step-by-step photos

Flooded lead acid batteries are used for stationary applications such as solar and wind powered electrical systems, and for mobile applications. If you need custom batteries of a specific size or output, wish to experiment with building batteries, or want to lower your costs by using recycled components and materials, *The Battery Builder's Guide* has the information you need.

Available in print from Amazon.com and for download in full color PDF ebook format at

www.batterybuildersguide.com

Download a free sample of The Battery Builders Guide in full color PDF at www.batterybuildersguide.com.

Titles from

Wheelock Mountain Publications:

by Phillip Hurley:

Solar II

Build Your Own Solar Panel

Build a Solar Hydrogen Fuel Cell System

Practical Hydrogen Systems

Build Your Own Fuel Cells

The Battery Builder's Guide

Solar Supercapacitor Applications

and also:

Solar Hydrogen Chronicles *edited by Walt Pyle*

Tesla: the Lost Inventions *by George Trinkaus*

Tesla Coil *by George Trinkaus*

Radio Tesla *by George Trinkaus*

Son of Tesla Coil *by George Trinkaus*

www.buildasolarpanel.com

Wheelock Mountain Publications
is an imprint of

Good Idea Creative Services
324 Minister Hill Road
Wheelock VT 05851
USA

17458184R00126

Printed in Great Britain
by Amazon